DEVELOPMENTS IN
GEOPHYSICAL EXPLORATION METHODS—4

CONTENTS OF VOLUMES 1 TO 3

DEVELOPMENTS IN GEOPHYSICAL EXPLORATION METHODS—4

Edited by

A. A. FITCH

Consultant, Formerly of Seismograph Service (England) Limited, Keston, Kent, UK

APPLIED SCIENCE PUBLISHERS LTD
LONDON and NEW YORK

APPLIED SCIENCE PUBLISHERS LTD
Ripple Road, Barking, Essex, England

Sole Distributor in the USA and Canada
ELSEVIER SCIENCE PUBLISHING CO., INC.
52 Vanderbilt Avenue, New York, NY 10017, USA

British Library Cataloguing in Publication Data

Developments in geophysical exploration methods.
—4.—(The Developments series)
1. Prospecting—Geophysical methods
I. Series
622′.15   TN269

ISBN-13: 978-94-009-6627-7   e-ISBN-13: 978-94-009-6625-3
DOI: 10.1007/ 978-94-009-6625-3

WITH 8 TABLES AND 94 ILLUSTRATIONS

© APPLIED SCIENCE PUBLISHERS LTD 1983

Softcover reprint of the hardcover 1st edition 1983

Photoset in Malta by Interprint Limited

PREFACE

Geophysical prospecting is an applied science and the range of scientific principles to be applied is very wide. In this collection of original papers, the application of many different principles is described in the search for sulphides, other metallic ores and radioactive deposits.

The papers are all concerned with surface observations and cover both the theory and the practice of the methods used. In all cases the advantages and disadvantages of the methods are described and their role in the detection of mineral deposits is discussed and placed in context.

Electromagnetic methods are covered in detail, involving the use of both electric and magnetic field effects. Techniques are described involving observations both at a number of discrete frequencies and with continuously changing frequency. In spite of the diversity of method it is interesting to note the strong links between the papers; two chapters, for example, start from the same fundamental illustration, first published by Won, of the basic relationship between source frequency, ground conductivity and depth of penetration.

The all-important economic aspects are not forgotten and the first chapter assesses the statistics of performance and describes their use in the shaping and management of an exploration programme.

The editor takes this opportunity to thank the busy men who have set aside time to write these contributions.

A. A. FITCH

CONTENTS

LIST OF CONTRIBUTORS

E. Gaucher

Edwin Gaucher & Associates Inc., 2406 Quatre-Bourgeois, Suite 200, Sainte-Foy, Quebec, Canada G1V 1W5

A. W. Howland-Rose

Scintrex Pty Ltd, 6 Tramore Place, Killarney Heights, New South Wales 20987, Australia

J. W. Motter

Whitney & Whitney, Inc., PO Box 11647, Suite 135, 1755 E. Plumb LN, Reno, Nevada 89510, USA

H. O. Seigel

Scintrex Ltd, 222 Snidercroft Road, Concord, Ontario, Canada L4K 1B5

W. M. Telford

Department of Mining and Metallurgical Engineering, McGill University, 3480 University Street, Montreal, Quebec, Canada H3A 2A7

I. J. WON

Department of Marine, Earth and Atmospheric Sciences, North Carolina State University, PO Box 5068, Raleigh, North Carolina, 27650, USA

Chapter 1

ESTIMATION OF SULPHIDE CONTENT OF A POTENTIAL OREBODY FROM SURFACE OBSERVATIONS AND ITS ROLE IN OPTIMISING EXPLORATION PROGRAMMES

E. GAUCHER

Edwin Gaucher & Associates Inc., Sainte-Foy, Quebec, Canada

SUMMARY

Geophysics can directly detect sulphide or graphitic orebodies containing gold, copper, zinc, lead, nickel, uranium or other metals, but for each anomaly corresponding to an orebody, geophysics also detects thousands of barren targets. This study is written with a single goal: to optimise exploration decisions in spite of all these false targets in order to find more ore with a finite exploration budget. After outlining the limitations of geophysics, the author reviews the practical signal-to-noise ratio and the cost effectiveness of the different geophysical methods used either singly or in combination in the search for sulphide orebodies. Many of the quantitative compilations on the actual effectiveness of geophysics come from a 16 man-year joint venture research project supported in the early 1970s by a group of four mining companies. The recommendations on the importance of target sampling (drilling) have already been published in an abbreviated form. They have today gained at least verbal acceptance by most explorationists. Other recommendations regarding non-conductive massive sulphide bodies or using single spacing, small separation induced polarisation surveys for gold exploration are new and have never been published before. Several of them are at variance with today's practice.

1. GENERAL CONSIDERATIONS

With one exception, this study will not explain how to do the surveys. In principle, we shall tend to assume that an area has been selected for geological reasons and that the geophysical surveys will be, or have been, performed.

1.1. Geophysical Parameters Used in Estimations

All mining geophysical methods provide information on the sulphide content of the ground investigated. In nature, sulphides and graphite are the only conductive minerals, and thus measurements of conductivity, be they pulse, EM (electromagnetic) or DC resistivity, inform us of their presence. Smaller, non-formational anomalies characterised by high conductivities are more likely to correspond to conductive bodies of economic sulphides. Such isolated anomalies are often targets for diamond drilling whereas the extensive formational conductors are often graphitic layers, especially in the Precambrian shield. Measurements of IP (induced polarisation) detect all sulphides, even those that are not conductive because they are too disseminated. Everything else being equal, the higher chargeabilities tend to correspond to greater sulphide concentrations. Because of their extra weight, sulphides can be directly detected by gravity surveys, especially when they occur in massive, near-surface orebodies. In many circumstances, gravity allows a direct quantitative estimate of the sulphide content. Finally, there is a higher probability of discovering sulphides under magnetic anomalies caused by magnetite or pyrrhotite than by random drilling, as the economic sulphides are frequently associated with these two minerals. The only method which does not provide direct information on sulphide content is refraction seismic, but it is used to interpret gravity anomalies.

The fact that some sulphide orebodies are associated with geophysical anomalies does not imply that all anomalies are orebodies, nor that such anomalies are reliable or even economic guides to orebodies. A review of the limitations of the different geophysical anomalies will be an important part of this study.

1.2. Precision Versus Cost of Sulphide Estimations

Before going into the subject in detail, we should like to warn the reader that, except for the results of some gravity surveys, and even then (as explained later) only if we have an independent knowledge of the presence of sulphide, the geophysical estimates are not accurate predictions but probabilities. Most of the time the conclusions will read like a guess, for example, 'Anomaly XYZ has one chance in 10 of containing a 3 m thick layer of massive sulphides', rather than a classical conclusion in a mining engineer's report: 'We have 95 chances in 100 of having 2 million tons ($\pm$0.1 million) of sulphides in a body 10 m wide ($\pm$1 m), etc.'

In principle, we should all like to achieve the second type of precision, but in practice, even when geophysics is able to achieve it, it is most likely not the recommended engineering approach. Exploration is an applied science and in most cases it is more efficient, once a target is defined, to check it by drilling to see if it contains economic concentrations of minerals, than to refine its nature by further and further geophysics. This consideration is overriding, and we should plan the yearly exploration budget so as first to reserve at least half of it for checking anomalies by drilling, and then do our best to select the targets to drill with the other 50% of the budget. Even if we do not control the budget, we can make our geophysical recommendations accordingly. Viewed in this light, this study will suggest how to use geophysical surveys at a reasonable cost in order to select the targets likely to contain the most sulphides. Basically, for each survey, we must think of how many targets to select and how much sulphide each target contains. The average cost of acquisition of each selected target should be equal to or lower than the cost of drilling or trenching it. Geophysics and drilling must proceed hand in hand for a better understanding and extrapolation of the geology and the potential economic value of an area. The whole process is one of engineering design: to find mines with the minimum expenditure, just like building a bridge without making it any more expensive than necessary.

1.3. Optimising the Exploration Budget

One can best demonstrate the necessity of properly allocating the exploration budget between drilling and screening by using the Monte Carlo simulation. Only two fundamental premises have to be made: (1) only the targets sampled by drilling or trenching can reward the explorationist; and (2) each exploration venture has a finite budget and the final cost of testing the average target is the total cost of the project divided by the number of targets tested. For a first exercise, let us set the cost of testing an extra target at $10 000 and the total budget for the venture at $100 000. Many geophysicists forget that a definite budget will not be exceeded, except if a discovery is made.

Let us assume that we have spent $50 000 so far to find 25 virgin and untested targets at a cost of $2 000 per target. We have two or three courses of action:

Option 1. We spend the rest of the budget to drill 5 out of the 25 targets. By doing so, we shall obtain at least 5 out of 25, or 20% of the odds to find a mine, and probably much more, as even with the limited information already available we are likely to select the five best targets among the 25 found.

Option 2. We decide to spend $30000 on further geophysics to select the best two of the 25 targets. We can now spend another $1 500 per target to upgrade them. However, as we are drilling only two instead of five targets, in order not to lose in the exercise, the two targets selected must contain on an average $2\frac{1}{2}$ times more sulphides than the five targets we would have drilled anyway had we followed the first option. The quality of the target must be upgraded by 250%, but the budget per target to do so has increased only from $2 000 to $3 200, or 55%. It is unlikely, to say the least, that at this late stage, a new geoscientific method can be found to upgrade the targets, at a rate twice as great as the first dollars managed to do, that could not have been applied a couple of months earlier!

Option 3. We hit a real procrastinator! Out of the two targets, he decides to select the best one! 'Let me save a drill hole, that's $10 000', he shouts. This wizard pretends to be able, for less than $10 000 of extra geophysics (otherwise there would be no saving), to select the only worthwhile target of the two. For less than the cost of a drill hole on top of the $80 000 already spent on geophysics, he will decide which of the 25 targets has nine times more odds of containing the mine than the second best of the 25 targets.

It may be considered superfluous to write out this scenario in detail, but in real life one major mining exploration company did spend $6 million (in 1982 dollars) while drilling only one single drill hole. Many companies spend as little as 5 or 10% of their 'moose pasture' budget on wildcat drilling, and very few of them go beyond 20%. This is one of the reasons for the higher rate of success of the small independent operations. Many exploration managers do believe that they spend a lot on drilling, but a rapid compilation of all the wildcat drill holes of the previous 12 months often proves how little was actually done: the big drilling effort is for next year!

Figure 1 illustrates the predicted cost of intersecting 1 m of sulphides by simulating on a computer the drilling of a selected number of conductors from a data bank of 158 targets. The cost of finding a metre

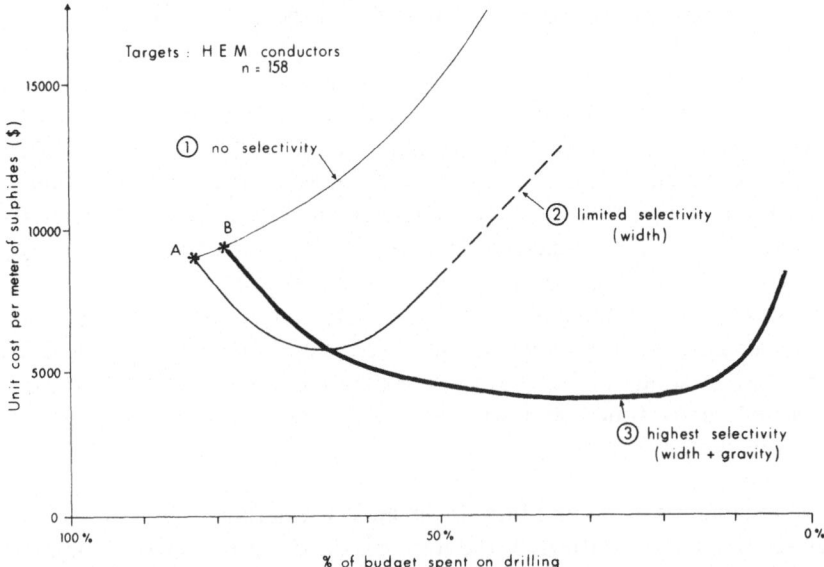

FIG. 1. Monte Carlo optimisation of exploration expenditures by selecting EM conductors for drilling at random, according to width or to gravity anomaly. For curve 3, the gravity anomaly was synthesised from the sulphides intersected, mixed with a random noise of an average amplitude of 0.06 mgal, typical for the Canadian shield.

of sulphides is plotted on the Y-axis, whereas along the X-axis we show the percentage of funds spent on drilling.

The data bank was created by compiling information on actual diamond drill holes (DDH) from exploration company files of wildcat ventures of several Canadian companies, the main ones being Soquem and Cominco. In the simulation illustrated, the cost of finding a conductor was assumed to be $2 000; the cost of drilling it, $10 000; and the cost of testing the conductor by gravity, $600.

If no selection criteria whatsoever were available, the lowest cost would be achieved by drilling all the targets, resulting in a cost of $9 000 per metre of sulphides at point A. One would then spend 83% of the budget on drilling! If by random choice some of the targets were left out, the cost would climb following curve 1, starting from point A.

If we classified the conductors according to their increasing width, rejecting the narrowest ones to begin with, the cost per foot of sulphides would start at A and then follow curve 2, first dropping to $6 000 per metre when spending between 60 and 70% of the budget on drilling, then rapidly climbing again as the drilling expenditures were further curtailed.

If we had also done gravity (or another efficient selection) at an extra cost of $600 per conductor, the cost curve would start at B at $9 500 per metre, then drop down to a broad minimum of $4 500 when spending on drilling between 60 and 15% of the total budget. There is no reduction in cost or gain in efficiency by drilling less than 50% of the budget. By drilling 50% of the budget rather than 15%, we shall find just as many massive sulphide deposits per dollar of expenditure and we shall also have a chance of finding all the other targets that do not react to gravity, such as gold or uranium deposits. The same conclusions would be obtained by simulating any other screening method. This simulation was repeated many times, and with any reasonable assumptions the conclusions were identical.

1.4. Fundamental Assumptions of Sulphide Estimations

In Section 1.2 we warned of the imprecision of estimations. Though we shall be dealing with each method, we shall further explain in this section the general limitations of surface estimates. Each geophysical survey is an indirect measurement of the content of sulphides, and the estimations are open to three types of error which decrease their precision. Fortunately, the third one is a compensating factor which helps to make the practice of geophysics more economically useful and valuable.

The first cause of error in any geophysical estimate is that a given quantity of sulphides can give widely different anomalies, depending on the grain size and dispersion of sulphides and on the shape and dip of the sulphide body. In Canada, for example, between one-half and three-quarters of the so-called 'massive' pyritic lenses and horizons are absolutely non-conductive whereas an extensive veinlet of pyrrhotite, only a few millimetres thick, gives an excellent conductor. In gravity surveys, horizontal or gently inclined layers of sulphides will give either no response or a weak, broad, questionable anomaly likely to be overlooked, whereas the same bed, steeply inclined, gives a definite signal.

The second cause of error is that substances other than economic sulphides give geophysical targets and responses. No one is really interested in finding graphitic layers, barren massive pyrite or pyrrhotite bodies. A massive sulphide body is, by our definition, a body of at least 200 000 tons containing 50% of sulphides by weight. In Canada, because of the regional grade of metamorphism of the Precambrian shield and of some of the Palaeozoic rocks in the Appalachian orogenic belt, most conductors that are found by airborne EM surveys are graphitic layers, metamorphosed from originally non-conductive carbonaceous or hydrocarbon-bearing strata. Many other conductors prove to be caused by conductive or weathered overburden, such as lake bottoms, glacial clays in Canada and Scandinavia, salty ground water in arid countries or simply deep weathering. Gravity anomalies are often caused by higher-density basic dikes or a ridge under the overburden.

The third 'compensating factor' that helps to make geophysics economically useful is the fact that the economic rewards are only related in the most general way to sulphide content. In nature, high sulphide content often corresponds to low economic rewards, but per contra, occasionally small sulphide or even graphitic concentrations correspond to exceedingly valuable gold or uranium orebodies, such as the rich uranium discoveries in the Athabasca basin (Paterson, 1979) or the Detour Lake gold discovery of Amoco in Ontario (Spiteri and Baria, 1981). What happens then is that some geophysical predictions, which may err as to the presence of sulphides, may still result in valuable orebodies, and even new mining districts.

1.5. Pertinent Recommendations After a Survey

The geophysical engineer writing recommendations resulting from a survey on a property must keep in mind the above three limitations of his predictions. Given a set of geophysical maps, he must avoid dogmatisms such as 'no massive sulphide on the property'. Instead, he should make a selection of the best available geophysical targets to be followed up by drilling or trenching, even if the anomalies are not perfect. He must keep in mind that drilling expenditures have to be approximately equal to the cost of geophysics. Furthermore, paradoxically, he should recommend additional geophysical surveys only if they will not only upgrade the targets already selected for drilling but also point to new and better drilling opportunities. As the odds are 99 or even 999 to 1 that no mine will be found in any case, it is easy after a survey to say that the area is no good, and anyone saying this has very little odds of being proved wrong. The effective geophysicist should have an understanding of the overall odds of exploration and keep in mind the necessity of balancing the expenditures of the investor (his employer) between screening by geophysics and drilling. A positive approach has resulted in many discoveries; a negative approach will never find a mine. Any survey not checked by drilling or trenching has 100% odds of being entirely wasted.

An understanding of the above is the single most important factor leading to proper exploration decisions as regards the selection of anomalies based on estimates of sulphide content. At the drilling stage, the decisions are of economic and engineering importance and not of scientific interest. The geophysicist, before finalising his report, must evaluate the cost of acquiring the selected geological area as well as the cost of performing the surveys that he is evaluating. He must base his recommendations not only on his estimate of the sulphide content of the targets, but also on their economic potential and cost. Let us develop an example. If there are only two targets after $50 000 of expenditure on a property in an area where a drill hole costs $10 000, then both targets must be sampled (drilled), regardless of whether they are 'good' or 'bad'. Actually, in some circumstances, the geophysicist should further review his results with the geologist, or review the geological information available himself so as to upgrade some of the other, weaker geophysical indications and select three other drilling targets. As we have assumed here that sampling an anomaly by a drill hole costs $10 000, a total of five drill holes should be spotted to balance the cost of expenditures to date. Of course, if the yearly budget authorised from the start was only $60 000, then $30 000 at most should have been spent on geophysics. Geophysical surveys should be followed by drilling within 12 months; otherwise, the funds invested in the surveys cannot benefit the investors.

2. ESTIMATING THE SULPHIDE CONTENT FROM A SINGLE SURVEY

2.1. Sulphide Content of Magnetic Anomalies

Some of the earliest successful applications of geophysics in the search for sulphide orebodies were obtained by drilling magnetic anomalies. For example, in Canada in 1945 the Quemont copper orebody was found by drilling a magnetic anomaly close to the Noranda mine. The magnetic anomaly was actually the orebody, the anomaly being caused by pyrrhotite and magnetite associated with chalcopyrite. By 1945 the conductive clay lenses had completely discredited DC resistivity surveys in the Noranda district and electromagnetic methods were just beginning in Canada, even though they were already being applied in Sweden.

The successive owners of the Quemont group of claims had already invested substantial funds in the direct search for ore. They had sunk a shaft, made drifts and cross-cuts, and drilled. Nothing having yet been found, a magnetic survey was requested at the beginning of another exploration campaign. One of the drill holes spotted on a magnetic anomaly intersected the orebody (Scott, 1948, p. 773). At least one other orebody was discovered in the same way in this district: the East Sullivan orebody discovered in Val d'Or the same year as the Quemont orebody (Koulomzine and Brossard, 1957, p. 177). The same year also saw the first discovery of a massive sulphide body under 8 m of moraine by an EM survey: the Macdonald orebody in Noranda (Boldy, 1979, p. 597).

Magnetism alone is seldom used in the direct selection of exploration targets for massive sulphides, mainly because magnetic anomalies are extremely common. As anomalies associated with orebodies are often weak, tens of thousands of similar-looking anomalies are caused by weak disseminations of magnetite (a fraction of 1%) in volcanic, intrusive and metamorphosed sedimentary rocks. Even if we restrict ourselves to the mining districts, magnetic anomalies are so common that, to our knowledge, no one has ever even tried to estimate their average sulphide content. Because of the low cost of ground and airborne magnetic surveys, they are used extensively as a first-order geological mapping tool. For example, almost all of Canada has been covered by airborne magnetic surveys, and a ground magnetic survey is done to cover a grid of lines whenever one is established for exploration. Magnetic maps improve the understanding of the geology, thereby indirectly suggesting targets for drilling. Magnetic anomalies are still drilled by some explorationists in the search for gold-bearing iron sulphide formations. In other cases, magnetic maps help to select areas for detailed investigation by electrical methods.

To summarise, in spite of their spectacular early success in mining exploration, magnetic maps today are used because of their indirect diagnostic value in the search for sulphides. Today, the Quemont and East Sullivan mines would have been found by EM surveys, and that is how we look for them now.

2.2. Finding and Selecting Conductors

Measurements of conductivity represent the single biggest expenditure in geophysical exploration for sulphides. Conductivity in natural materials varies by over ten orders of magnitude, from the high resistivity of dry, deep rocks or rock salt (up to 10^6 Ωm) to the high metallic conductivity of natural silver, gold or copper (10^{-8} Ωm). It was early recognised that many massive sulphide bodies containing valuable base metals and gold are often very good conductors.

Conductivity surveys are therefore performed to find massive sulphide bodies hidden under the surface. The methods used for these surveys since the 1920s have been and are constantly evolving. Geophysicists have always attempted to optimise the information provided by conductivity surveys. The earliest maps were DC (direct current) resistivity. They were followed by electromagnetic surveys which first measured the inclination (deformation) of the primary field at one or several frequencies, then measured its intensity and phase angle. Today, pulse surveys penetrate even deeper, under more conductive overburden, and no one uses DC resistivity surveys any more except in conjunction with induced polarisation investigations. Ward (1979) gives an excellent résumé of all the EM methods, and Crone (1979) gives examples of the early successes of pulse EM.

After the value of ground electromagnetic surveys was recognised, they became airborne in the 1950s in Canada. Pulse surveys, because of their complex electronics, were used in the 1960s only in the airborne mode before being adapted for ground surveys, where they are now used on a large scale.

This constant search for improved resolution is directly related to the original observation that in mining districts, when many near-surface materials are conductive, the more conductive materials are more likely to be sulphides in the bedrock. We shall prove in this study that the situation is more complex. Let us review what is available in terms of measurement methods.

2.2.1. DC Resistivity Surveys
The early DC resistivity surveys, from the mid-1920s on, were successfully used on a large scale in areas of high surface ground conductivity, such as the moraine-covered areas of parts of Sweden and of the Chibougamau district, in the centre of the Province of Quebec. Because of favourable local conditions, DC resistivity was so successful in Sweden that even recently some Swedish geophysicists did not believe in the usefulness of IP measurements that can be done simultaneously (Parasnis, 1975, p. 225). On the other hand, DC resistivity fails hopelessly even in areas of moderately conductive lacustrine clays, such as in the Noranda district, where humorous poems were written in the 1930s about the geophysicist finding a mine under every swamp. The reason is that in DC surveys, only the apparent resistivity can be measured, and there is no difference between the anomaly of a lens of lacustrine glacial varved clay and that of an orebody.

2.2.2. Dip Angle Electromagnetic Surveys

In the late 1930s dip angle electromagnetic surveys started measuring deformations of a primary field by measuring the field's inclination. Dip angle surveys can be made with a great multiplicity of configurations and frequencies, using fixed or mobile transmitters, or even transceivers. By using two or more frequencies, they can discriminate between 'poor' and 'good' electrical conductors. They have been and still are being used extensively; even in clay areas in the Canadian shield, they can screen out the poorer conductors which are often but not always caused by surface overburden. By using a low frequency, dip angle surveys can trace good bedrock conductors even under a moderately conductive surface layer.

One of the more widely used dip angle instruments, the VLF, uses as transmitters the US Navy's 16 000-cycle radio communication stations for submarines. The frequency is so high that it does not discriminate at all. In any area with some conductive overburden, the great number of conductors that the VLF discovers every time the overburden cover changes thickness gives geologists searching for gold a great opportunity to use their intuition in selecting the most favourable drill targets.

Because of the high calibre and the good luck of the geologists, the method is credited with several discoveries of gold-bearing orebodies and veins in the Timmins–Val d'Or area, many of which do not behave as 'conductors' in any other conductivity survey. An IP anomaly is often not an electrical conductor, but a VLF anomaly must be caused by a conductor, be it bedrock or overburden. The mystery could probably be explained if the VLF survey was repeated after the clay overburden was removed before mining. It would probably be observed that the anomaly had disappeared, just as it does not exist over the outcropping portion of some of these 'VLF' discoveries, such as the Silverstack orebody (Gaucher, 1981).

Actually, even a good conductor, caused for example by a lens of massive sulphides, is not detected by a VLF survey if the conductor happens to be covered by a layer of conductive clay. On the other hand, a VLF instrument is an extraordinarily efficient tool in the hands of a prospector or geologist who, while doing the survey, can judge whether the conductor lies under clay in a swamp or under the moraine on an elevation; in the latter case, it can be a mine like the Lessard copper discovery in the Frotet district of Quebec (Reed, 1979, p. 633).

2.2.3. Horizontal Loop and Other Related Methods

We have grouped together in this section all the electromagnetic me-

thods that under most conditions give an accurate estimate of the
conductivity–width of an anomaly, either by measurements of phase and
quadrature, or quadrature at multiple frequencies. These instruments are
called Slingram, Turan, MaxMin, etc. They are also available in airborne
configurations, either for helicopter or aeroplane. The conductivity
estimations are reasonably independent of the shape and the dip of the
conductive body and, with the horizontal loop surveys, a realistic estimate
of the width of the conductor can be made from the width of the
anomaly. Because of the abundance of accumulated data, we have been
able to compile data banks which allow us to make predictions concerning the sulphide content of conductors according to their width and
conductivity.

Because of the slightly higher cost and lower depth penetration of
horizontal loop surveys, and in spite of the more quantitative conductivity estimates and a better resolution of short, near-surface anomalies, the HEM surveys did not really replace the dip angle surveys. Both
methods coexisted side by side, and only when improved electronics
allowed the use of wider separations and greater depth penetration did
the horizontal loop measurements become prevalent. Even today, with
the third generation of pulse surveys coming into its own, the two older
methods coexist.

2.2.4. Pulse Measurements

This method measures the same eddy currents as the continuous-wave
electromagnetic systems, but here the transmitter sends a pulse and the
receiver measures the decay of the eddy current after the transmitter has
been cut off. Complex electronics sample a number of time windows
which for the first time allow the separation of the effects of conductivity
of the sulphide body from the geometric effects of its shape. Pulse surveys
can also be done in drill holes allowing the measurement *in situ* of
conductivities of orebodies. As more and more surveys are done by pulse,
it will be possible in the future to compile much better data banks on
conductivity and sulphide content of orebodies and eventually, to improve the surface prediction of sulphide content. At the present time,
only phase-angle compilations are available.

2.3. Sulphide Content of Conductors

Because of the overwhelming effect of grain size and impurities, past
experience, which can be expressed in compilations, is the best guide to a
quantitative sulphide estimation. Three types of compilation will be

described. The first is a systematic sampling of the quantity of sulphides encountered in the conductors drilled in a large area of Quebec; the second quantified the sulphides encountered in conductors according to their conductivity–thickness product, as determined from real and imaginary components. The third compilation was done according to the apparent thickness of the conductors, as this factor was extensively used to select anomalies found with horizontal loop surveys. In such surveys, the thickness of the conductor was estimated by subtracting the length of the cable from the width of the horizontal loop anomalies; empirically, the thicker the conductor, the greater the likelihood of substantial sulphides. All these compilations were made in 1971–72 at Soquem by a group of geoscientists headed by the writer (Gaucher *et al.*, 1972). Half the funds were provided by Soquem and half by New Jersey Zinc, Cominco, Pennaroya and the Geological Survey of Canada.

2.3.1. Sulphide Content of an Average Conductor

Figure 2 is a compilation of the quantity of sulphides reported in some 70 diamond drill holes drilled across conductors in the Precambrian

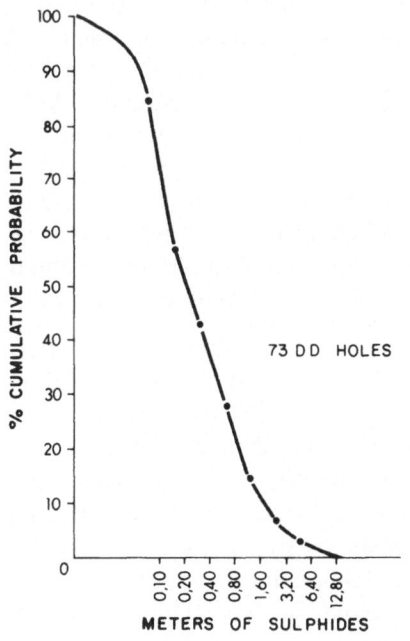

FIG. 2. Sulphide content of 73 diamond drill holes drilled to test EM conductors. This cumulative histogram expresses the probability of finding more than a specific thickness (in metres) of sulphides in any EM anomaly without regard to its geophysical parameters. Note the logarithmic scale on the horizontal axis. (From Quebec Department of Mines, files of mining companies' exploration campaigns.)

shield (Gaucher and Nadeau, 1972, p. 4). The drill holes were compiled from the statutory reports of the Ministère de l'Energie et des Ressources, filed by mining companies in an area measuring 100 by 200 miles in northern Quebec; the study sample was constituted by total sampling of the same 2 square miles of every 100-square-mile township. The area is partly underlain by volcanic rocks which had been explored for conductors by airborne EM surveys, followed by ground EM surveys and then by drilling by several different mining companies. Each drill hole reported in a given square mile was examined as to why it was put down; if it was drilled on a conductor, we estimated the content of sulphides in the drill hole from the geological log. The content of sulphides was compiled as equivalent massive sulphides in feet or metres; for example, 20 m of 50% of sulphides will be reported as 10 m of sulphides. Minor disseminated sulphides (1–2%) are not compiled except when in veinlets. The same method of quantification of sulphides is used throughout this paper. When present, economic sulphides were also noted.

Though the conductivity–thickness product of the conductors compiled was often unknown, it is likely to have been often high (from 10 to 100 mho). Crude as the methods of selecting drilling targets were, they were effective in screening out the best conductors in any given area, although they also strongly tended to bias for the conductors close to the surface.

No correction was made for the angle of intersection of the drill holes with the target, but any overestimation is not likely to have been high, as most drill holes tend to flatten out, and as they were put down from the downdip side of steeply dipping beds. Almost all the geological logs of the drill holes reported intersecting a conductor but, from my experience, this is likely to be a case of self-fulfilling expectations. Even assuming that 10% of the drill holes are misplaced, and allowing for many other inaccuracies, the conclusions are overwhelming:

(a) The median drill hole that intersected a conductor contained a total of less than 0.3 m of sulphides.

(b) 90% of the conductors contain less than 1.5 m of sulphides.

(c) Most of the conductors are associated with and caused by graphitic layers, and the sulphides reported are pods of pyrite. The second most common cause is narrow pyrrhotite veinlets, which often add up to less than 0.1 m of sulphides in a DDH.

(d) Much of the sulphides was found in less than 5% of the conductors.

(e) No economic mineralisation whatsoever was reported in any of the 74 conductors.

Actually, our estimate of the frequency of economic mineralisation occurrences in such types of blind systematic endeavours would be as low as one in each 1000 targets investigated in a Precambrian-shield-type environment. Assuming a 1982 cost of $15 000 per target, half for geophysics and half for drilling, it would cost up to $15 million to intersect the first worthwhile target, and perhaps several times that much to find the first mine. This is not the most rejoicing projection for a mine seeker, and any method that could improve this performance would be most welcome.

Even if it is only a 'guesstimate', the high cost of discovering a mine by systematically flying airborne surveys and drilling conductors should not come as a surprise. It is only objectively stated here, but it may not be different from the average real cost of any discovery even in old-established mining districts. This cost is probably lower only during a mining boom, when a new mining district is being developed and when several additional orebodies are found in the same mineralised environment. Prospecting of new districts with new methods and ideas also sometimes reduces this cost for a while.

2.3.2. Sulphide Content According to Characteristics of Conductors

For this aspect of the study, Gaucher et al. (1972) used data banks constituted from results of exploration files of New Jersey Zinc, Cominco and Soquem as well as reports filed by companies for assessment. Conductivity–width and width of a conductor are estimated separately from HEM surveys. Other geophysical parameters were also compiled when available. A simple histogram was sufficient to represent the sulphides found in a number of conductors in the data bank. To represent the sulphide content in conductors of different conductivity–widths, the simplest representation is a superposition of histograms as shown in Fig. 3. The conclusion is clear: in a data bank of 124 conductors, the better conductors, those whose conductivity–thickness is higher than 20 mho, contain more sulphides.

To quantify this statement, we must first define a 'desirable target'. In exploration, only very few targets are desirable and pay off; these are the good targets. It is reasonable to assume that a most desirable target contains 8 m of massive sulphides. Among the 41 good conductors ($\sigma > 20$ mho), 4 contain more than 8 m of massive sulphides, whereas among the 25 poor conductors ($\sigma < 5$ mho), only one contains that amount

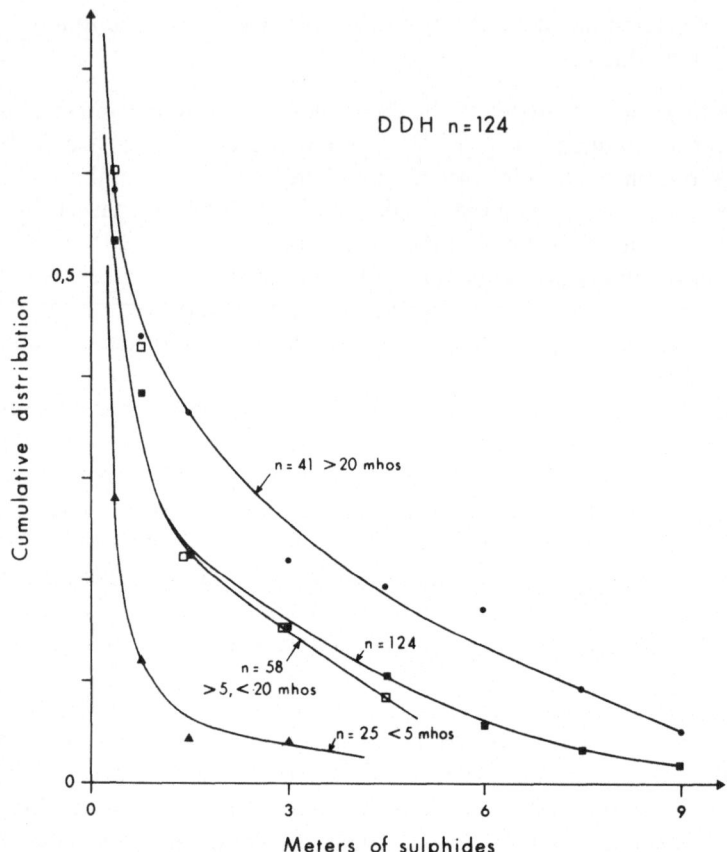

FIG. 3. Sulphide content of 124 diamond drill holes classified according to the conductivity of the EM conductors tested. This cumulative histogram shows the fraction of the total number of DDH (vertical scale) containing more than a certain quantity of sulphides (horizontal scale). Crosses represent all 124 drill holes made during the regional exploration campaigns of a number of companies in the Canadian shield, whereas dots, squares and triangles represent those drill holes which tested either good, medium or poor conductors. The graph illustrates that the conductors that have a higher conductivity–width contain more sulphides than others. n = number of DDH compiled for the curve shown.

of sulphides. We may conclude that the 'good' conductors are 2.5 times better than the 'poor' ones. However, we shall see that reality is more complex.

To represent simultaneously the quantity of sulphides in terms of two factors, a more complex representation is necessary. In Fig. 4 we have

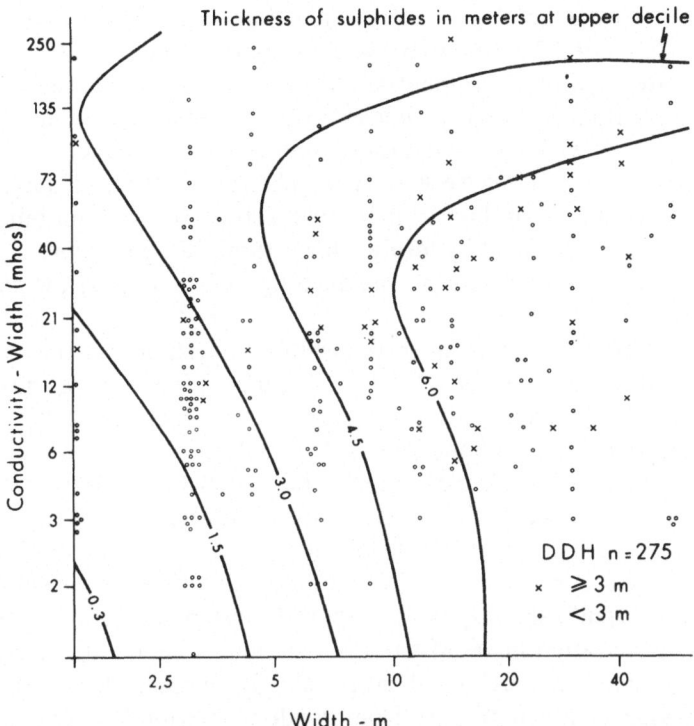

FIG. 4. Sulphide content of 275 diamond drill holes plotted according to the width and the conductivity–width product of the EM conductor tested. The graph illustrates the sulphide content of each of the 275 DDH compiled in a computer data bank, as classified according to their width (horizontal scale) or their conductivity (vertical scale). The DDH were drilled to check EM targets during regional exploration campaigns. DDH which intersected less than 3 m of total sulphides are shown as dots; crosses indicate holes with more than 3 m. The contours represent the sulphide content at the upper decile, computed with a three-dimensional probability estimate, from the sulphides reported in each drill hole. Wider conductors (horizontal scale) correlate well with more sulphides, whereas conductivity–width (vertical scale) barely correlates with sulphide content. Astonishingly, EM conductors that are at the same time very good and very wide may contain fewer sulphides. We found the upper decile to be more significant than the mean, as most drill holes contain hardly any sulphides.

plotted the conductivity–width versus the width of 275 conductors from our data bank, all of which were drilled during regional exploration programmes. The vertical scale is the conductivity in mhos and the horizontal scale is the width in metres. Both scales are logarithmic to ensure a more uniform distribution of the drill holes. Each drill hole is represented by a dot or by a cross if it intersected more than 3 m of massive sulphides. To represent the quantity of sulphides, we decided to contour the quantity of sulphides intersected in the best decile (10%) of the holes within a moving window which selects the rolling sample referred to below. Thus the contours represent a probability for a given conductivity and width: we have a 0.1 chance of hitting a quantity of sulphides equal to or higher than that shown by the contours. The contours were calculated by fitting a logarithmic distribution to a rolling sample of drill holes shown on the diagram and by finding the value of the upper decile.

To the writer, the diagram proves without question that the width of the conductors as given by HEM or Slingram surveys is an important parameter for predicting the sulphides in a conductor; width is much more important than conductivity as the contours are subparallel to the conductivity axis. Wide, medium-to-good conductors (10–100 mho) are only marginally better than wide, poor conductors. Only in narrow conductors does conductivity help. Surprisingly, we can observe that wide, excellent conductors (> 100 mho) may have a lower probability of containing significant sulphides than wide but poorer conductors. To increase penetration in HEM surveys, ever longer cables are being used. Any two good narrow conductors, closely spaced, then appear as pseudo-wide conductors. Just from random distribution, two narrow conductors are more likely to appear as a single extra-wide conductor than as just a wide conductor. Why the same does not apply to poor conductors is still a mystery.

Selecting wide HEM conductors as drilling targets for massive sulphides has been practised by many users of HEM surveys. It might be one of the reasons why, in Canada, some users of HEM (e.g. Hudson Bay Mining) have traditionally been more successful in their regional exploration than others, as users of vertical loop EM do not routinely measure widths of their conductors.

This observation, if accepted, may help to keep the popularity of horizontal loop surveys at a high level, as they allow us to estimate the width of conductors easily and accurately. Finally, it is important to note that the conductivity alone, as opposed to conductivity–width product (mhos), is likely to have only a weak correlation with the quantity of sulphides.

The favourable quality of wide but poor conductors leads to an interesting extrapolation as many massive or submassive pyrite bodies are absolutely non-conductive and can be electrically detected only by IP surveys. Such non-conductive pyrite concentrations may contain economically significant zinc, silver and gold concentrations. Thus, wide bodies of sulphides may always be favourable targets except that sometimes they can be detected because of their conductivity and at other times only by IP. As we are later going to suggest, they can also be detected by gravity surveys (Brock, 1973).

Poorly conductive or non-conductive massive sulphide bodies probably constitute many near-surface and worthwhile targets in the old mining districts already prospected for conductors by surface methods. So far, such orebodies have been discovered almost accidentally (Reed, 1981).

2.4. Sulphide Content of IP Anomalies

To estimate properly the sulphide content of IP anomalies, two parameters are necessary: first, a proper understanding of how to interpret and quantify IP responses; and second, some probability functions derived from compilations of sulphide content of drill holes across IP anomalies. As far as we know, no such compilation has yet been done, and for good reasons, as only a few people today properly attempt to quantify an IP response in terms of sulphides. We shall propose here a 'sulphide factor' to be used to measure and quantify IP responses. As it is likely that the sulphide content of IP anomalies will be proved in future to correlate closely to the sulphide factor, we shall specify how to measure, then how to calculate, the sulphide factor. The discussion is limited to the case where surface resistivities are lower than deeper bedrock, which is to be investigated.

2.4.1. Selecting the Proper IP Array

Any IP survey is made with two current electrodes which inject the current into the ground, and two potential electrodes which measure the deformation of the signal by eventual sulphides. Proper IP electrode configurations are a must for estimating sulphide content and, to qualify, all four electrodes should move together closely spaced over the ground to be investigated. Dipole–dipole and Werner arrays are good.

Gradient array IP surveys, which may be somewhat cheaper per mile, are not very useful in estimating sulphide content. When modelled in the laboratory, gradient array surveys give recognisable anomalies, but in the field the results are often broad chargeability highs caused by any extensive pyritic horizon present. The highs themselves are distorted by variations in the overburden surface conductivity. No satisfactory estimate is obtained for overburden conductivities. Presented with the resulting maps, the geophysicist examines the chargeability profiles and tries to select a number of slope changes for detail by a dipole–dipole array. A single spacing dipole–dipole is almost as inexpensive and gives a much better map, directly quantifiable in terms of sulphide content. Even in the Pine Point lead–zinc district, where overburden conditions are exceptionally uniform, Lajoie and Klein (1979, p. 658) report that gradient surveys are likely to miss near-surface orebodies unless closely spaced current electrodes are used, and even then the anomalies are much better with a pole–dipole survey. Gradient surveys have found mines, just as magnetic anomalies have found them, but today other methods are more cost-efficient.

Pole–dipole IP surveys are much more useful than gradient arrays, but their value is reduced if the fixed current electrode happens to be planted near a body of sulphides; all the readings then become anomalies. This is twice as likely to happen, of course, in gradient surveys.

2.4.2. Selecting the Proper Electrode Spacing

During IP surveys, geophysicists have always measured resistivity and chargeability simultaneously. Indeed, in any frequency survey the chargeability is derived from resistivity measurements. Probably for this reason, it is believed that to measure the chargeability (or frequency effect) of a formation under conductive overburden properly, widely spaced electrodes have to be used, just as they are used to estimate the resistivity in a Schlumberger array by spacing the current electrodes further and further apart. Furthermore, geophysicists had (and many still have) the impression that unless they spaced these electrodes far enough apart in order to measure the higher bedrock resistivities, their electrode arrays were not penetrating the overburden and thus the survey was not detecting the sulphides underneath.

Such reasoning is erroneous. Chargeability (or frequency effect) is a pulse phenomenon and, with an accurate instrument, one can measure chargeabilities of the bedrock across a significant thickness of conductive overburden, even with closely spaced electrodes. The absolute value of the chargeability is lower at smaller separations when measuring across a layer of conductive overburden, but the signal-to-noise ratio of an anomaly is better.

The 1972 survey in the vicinity of the Phelps Dodge orebody in Quebec's La Gauchetière township is the best example of such mistaken reasoning (company files, Noranda Exploration). The orebody is covered by 70 ft of clay and till. The profiles over the orebody were surveyed with dipole–dipole arrays of three sizes (a) at four multiple separations (n) with a frequency-domain IP. The survey was first done with a 400-ft dipole ($n = 1 - 4$), then with a 200-ft dipole ($n = 1 - 4$) and finally with a 100-ft dipole ($n = 1 - 4$). The orebody was well detected only on the first separation with a 100-ft dipole, and weakly at the first separation of the 200-ft dipole (and the equivalent second separation of the 100-ft dipole). Neglecting these observations, the geophysicist followed the established opinion and concluded in his report that because of the low surface resistivities (70–100 Ωm) he was not penetrating the overburden. He recommended a broad and wide 400-ft ($n = 1 - 4$) survey of the whole property. He did not find any new targets.

Eight years later the measurements were repeated with time-domain equipment with arrays of $a = 100$, 200 and 300 ft and the same results were obtained. Interestingly enough, on the first separation of the 100-ft survey, the signal-to-noise ratio of the anomaly was comparable in either time-domain or frequency-domain surveys. Both instruments were of the same vintage (1970) and it would be interesting to repeat the measurements to find out which method would come out ahead with today's more accurate equipment.

The same improvement in the quality of the IP response with smaller separations, despite the apparent lack of penetration of the overburden, can be seen over the Mogador orebody, over the Thompson Bousquet orebody, and in many other surveys not illustrated. In spite of the overwhelming evidence, old beliefs are hard to change; recently, some authors even recommended measuring only second and third separations. Luckily, with today's simultaneous multi-separation measurements, geophysicists tend to reduce their first separation arrays, thus often obtaining useful results in spite of the theories.

The phenomenon is exactly reproduced in computer or laboratory modelling, but as the overburden is uniform in the laboratory, one does not notice the dramatic fall-off of the signal-to-noise ratio that one observes when greater separations during actual field surveys are used. Geophysicists tend to underestimate the havoc created at wider separations in the apparent resistivity measurements by irregularities in the overburden cover. For example, in a resistivity survey, every outcrop sticking out of the conductive glacial clay layer appears on a multi-separation dipole–dipole survey as overlying a superconductor! The havoc wreaked on a gradient survey is even worse.

To summarise, we recommend running IP surveys with a dipole equal to the average expected depth of overburden, and if three or more separations are done, a dipole of half the depth of overburden can be used. In practice, we suggest dipoles of 25 m or 100 ft. A 50 m dipole can be used if an inexpensive single separation survey is required. However, smaller targets, such as gold-bearing sulphides, could be missed with the greater separation. In no cases seen by the writer does the gradient array give a better signal-to-noise ratio than a dipole–dipole, so that the gradient in practice should logically not have a greater penetration under any condition. A moving Werner array could probably be run just as fast and could even give better results.

2.4.3. Calculating the Sulphide Factor
The sulphide factor (SF) would be expressed in the same units as the metal factor used for many years by Halloff:

$$SF = \frac{Chargeability \times 2000}{6.6 \times \rho_a} = \frac{FE \times 2000}{\rho_a}$$

where ρ_a = apparent resistivity in $\Omega\,m$, FE = frequency over one decade, and chargeability is expressed in ms relative to M331. We assume standardised FE and chargeability as per Sumner (1979, p. 126).

We propose a new name only because we recommend that the SF be calculated for all dipole–dipole surveys, be they time domain or frequency domain, and only at the first separation. When overburden changes in character laterally, calculating this sulphide factor for higher separations reduces its usefulness and results in erroneous anomalies and decisions, as over the Phelps Dodge orebody (Gaucher, 1980).

From compilation of sulphide content associated with a certain sulphide factor and by standardising and calibrating the sulphide factors, one will probably convert the SF into a useful index of sulphide content. The calibration may have to be done on an upper decile just as for the width of the conductors. Snyder and Merkel (1977), as reported by Sumner (1979, p. 125), have started to do so, but more work is probably needed.

In the meantime, drafting contours of the sulphide factor from properties allows consistent mapping of sulphide concentrations and layers, even under widely variable overburden, as shown on the many maps made by our group for mining companies.

2.4.4. Predicting Sulphide Content of an IP Anomaly
Until proper statistics are gathered on IP anomalies, we recommend

measuring their width and using it as a parameter in Fig. 4 to predict their sulphide content. We believe that they are likely to contain as much sulphide as EM conductors of the same width, as we have proved to our satisfaction that width and not conductivity is the best criterion in conductors. If a width cannot be estimated, then the anomaly can be assumed to be a conductor of unknown width. As already mentioned, several large sulphide bodies, up to 300 m in width, have been outlined by IP, and later by gravity, for example in the Yukon and in Quebec. Rich zinc, copper or gold orebodies in these sulphide concentrations have then been found by drilling (personal experience).

2.5. Sulphide Content of Gravity Anomalies

The concentration of heavy sulphides creates gravity anomalies. Their excess mass causes local increases in gravity attraction which can be accurately measured with today's instruments. The station elevations can be measured easily and accurately at negligible cost, even in heavy bush, with electronic hydrostatic levels (Hood, 1980, p. 63). Computers provide easy and inexpensive data reduction terrain corrections, even right in the field.

The expected actual near-surface mass (M) in tons can easily be approximated by the following formula (Parasnis, 1975, p. 271):

$$M = 23.9 \frac{\rho_2}{\rho_2 - \rho_1} \Sigma(\Delta g \times \Delta S)$$

where ρ_1 and ρ_2 are the densities of rock and ore, respectively, Δg is the average gravity anomaly (mgal) considered and ΔS is a small element (m^2) of the area of measurements.

The total cost of gravity per mile or kilometre is comparable to that of an IP or a pulse survey. In spite of its advantages, gravity is not used extensively by many companies, and seldom by itself. We believe that it will become more popular when its possibilities and its shortcomings become better understood. We shall attempt to explain them here.

2.5.1. Shortcomings of Gravity Surveys

The rather limited use of gravity may be explained by the following shortcomings, problems, and perhaps misunderstandings, of the capacities of the method:

(a) Gravity does not detect flat-lying or gently dipping sulphide layers which are easily detected by electrical prospecting. Fat, flat-lying lenses can be detected.

(b)   Gravity reacts only to important masses of sulphides: a 3 m thick layer is near the minimum size, if it is steeply inclined. Some economic orebodies occur associated with less than 3 m of massive sulphides.

(c)   As large masses of sulphides are very rare, a gravity survey very seldom results in unquestionable targets of the type illustrated in Fig. 6.

(d)   The resulting cost for a quality target is high. Soquem, for example, encountered only one exceptional anomaly, a barren pyrite body greater than 1 mgal (1.8 mgal) in ten years, 20 000 gravity stations and some half a million dollars of expenditures. Some ten smaller bodies, all barren, were also detected during that time, but many of the poorer targets drilled were not sulphides, creating a strong feeling of frustration among the geologists.

(e)   The numerous targets of electrical prospecting give geological information and allow the geologists to select the targets according to their geological intuition. Local gravity surveys seldom provide much information on geology.

(f)   The drill holes on the numerous targets of electrical prospecting provide interesting geological information on the distribution of sulphides, even when they are weakly mineralised. False gravity anomalies are usually totally sterile.

(g)   In areas most active in exploration, variations in the nature and thickness of surface materials, especially clay and till in glaciated areas, increase the gravity noise, leading to numerous false anomalies.

(h)   Gravity cannot find even large sulphide targets buried deeper than 200 m, and 100–150 m is the upper practical depth for medium-sized orebodies. Even if the gravity anomalies of deeper orebodies are perceived and measured, the resulting wide gravity bumps are not diagnostic and cannot be distinguished from geological contacts and background noise.

Let us now examine how gravity can be used, despite its limitations.

2.5.2. Predicting Sulphides by Gravity

To estimate not only the size of the body but also the probability of its being a body of sulphides, one has to estimate the 'noise' generated in gravity surveys by other causes, at the wavelengths that can correspond to near-surface orebodies.

A compilation of the pseudo-anomalies, or noises, observed in surveys totalling 6000 gravity stations, was made and is shown in Fig. 5. The stations were spaced every 30 m along lines at right angles to steeply dipping formations of the Canadian shield. The milligals express the average height of the anomalies above the mean gravity value over a window (cord) of given length, as calculated by a computer which processed all the gravity measurements. Because of numerous drill holes and other information from electromagnetic surveys, we believe that these anomalies are not caused by sulphides.

For example, if we measure from a 180 m window and a probability of 2σ, the graphs express the observation that when we do a gravity survey along a line with stations every 30 m, we shall find an anomaly of an average height of 0.17 mgal on the average of every 40 stations, or for about every 1.2 km surveyed. The peak of such an anomaly is likely to be greater than 0.3 mgal, similar to Selco's Uchi Lake orebody in Ontario,

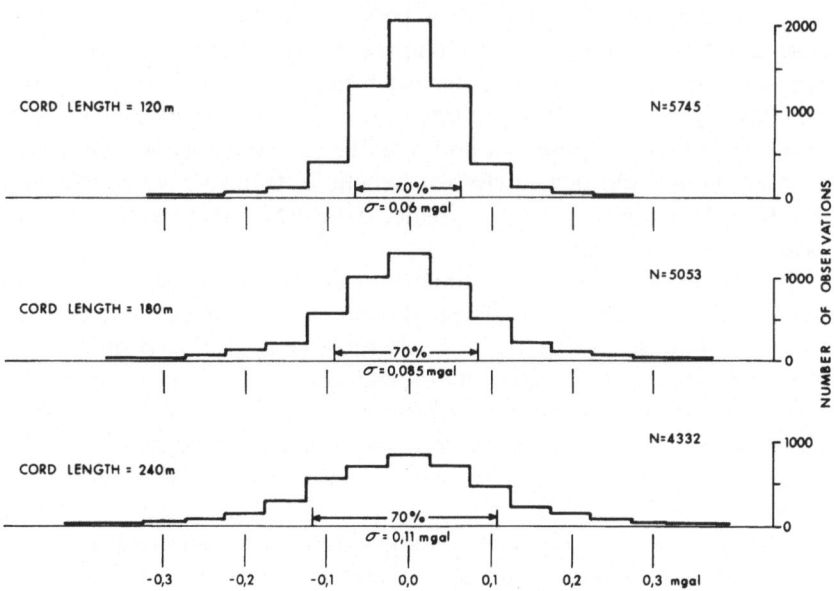

FIG. 5. Histograms of gravimetric noise compiled from surveys in the Canadian shield. N is the total number of observations, σ is the standard deviation. The noise is averaged for a certain window length; noise peaks are twice as high as the mean level of noise above the average gravity value in the window. (Courtesy: Soquem.)

Canada, which contained about 10 m of massive sulphides. The expected noise and the Uchi Lake orebodies anomaly are shown in Fig. 6.

Because of this random noise when we do a systematic gravity survey, we are sure to find, for every 1.2 km of survey, an anomaly equal to or greater than 0.3 mgal, at a cost of $1000 per anomaly. The odds of hitting 10 m of even barren massive sulphides at such a randomly generated spot are certainly less than one in a thousand (1:1000). Such a gravity high is a much poorer target than any EM conductor!

Even with inexpensive drilling the exercise would be rather depressing, and 90% of the budget would be spent on drilling. Systematic gravity can be used if we increase the minimum size of the target to be checked to 3σ, or about 0.4 or 0.5 mgal. With bigger targets we could use 50 m station spacings and find an anomaly every 400 gravity stations. Thus, we would probably spend 50% of the budget on drilling. Pennaroya in 1968 went into exploration in the Canadian shield with exactly that approach in an area of extensive overburden, where the electrical methods did not perform well. They covered only half of their prospecting perimeter before abandoning, but had they persevered, doubled the area and quadrupled their budget to $2 million, they would have discovered the huge Detour massive sulphide deposit which was found ten years later only slightly north of their projected gravity perimeter by Selco (Reed, 1981). The Detour deposit is a poor conductor and it was only perceived by later, more sophisticated electromagnetic instruments. As Pennaroya, or rather Pierre de Bretizel, had predicted, his EM instruments might not have picked it up at all.

Gravity noise in areas of uniform thin till, such as in areas of Newfoundland or some non-glaciated areas, is only half as great as in the areas we sampled for our study. Favourable geological environments in these areas can be covered more efficiently with gravity, even for medium-sized targets. Because of lower gravity noise in many other areas, gravity is likely to be the best tool to use in the search for pyritic copper orebodies, such as those near ancient copper workings in Cyprus or Portugal.

Even if the average sulphides actually found in each anomaly are much smaller than predicted from the size of the anomalies, the result may still be much superior to random sampling by drilling favourable areas or spending the funds on non-specific preliminary studies.

To summarise, gravity used alone may be a good tool to find sulphide bodies and to give a quantitative estimate of their size. The cost limitations suggest its use in old proven mining districts for large pyritic

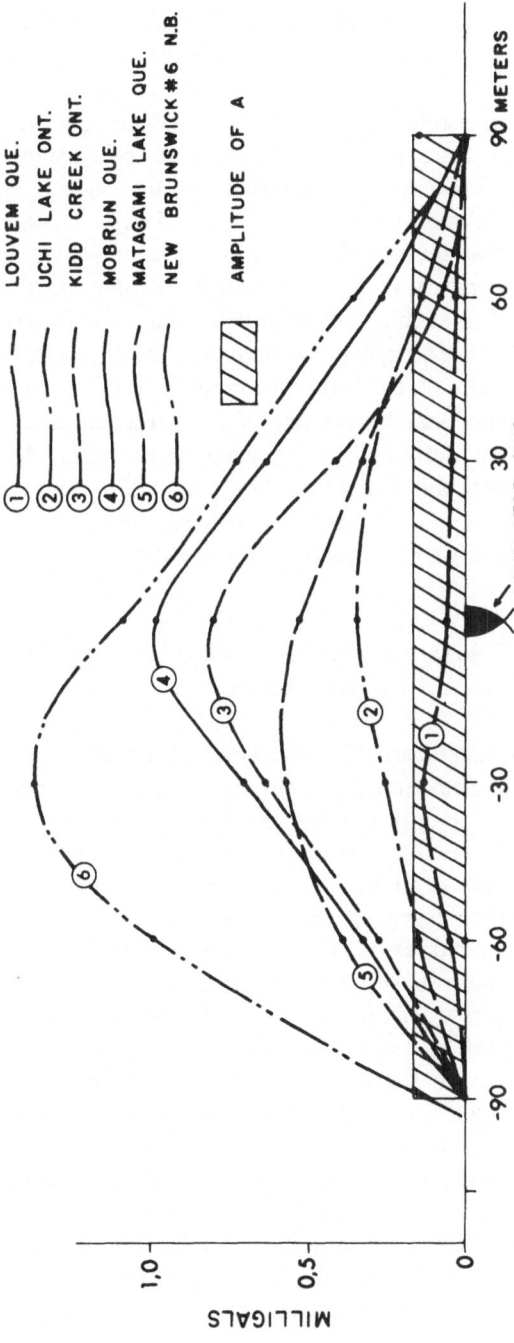

FIG. 6. Gravity anomalies over six massive sulphide orebodies in Canada. None of these mines, except for Mobrun which is not an orebody, was discovered by gravity. The dashed area represents the noise observed in one out of 40 stations (2σ) in the shield. Gravity alone could have discovered the bigger orebodies and it could have helped to select the right target for the smaller ones. During the Soquem study, gravity surveys were done or were recompiled over some 25 Canadian mines, and the six curves illustrated are representative of the range of responses observed.

lenses, which often carry valuable base and precious metals. Gravity also excels in predicting sulphides of pre-existing electromagnetic targets, as explained later.

Today it is impossible to predict the sulphide content of a given anomaly found by a gravity survey because of the existence of gravity noise, and it will remain so until compilations are made in specific areas of the frequency of the occurrence of sulphide bodies. Thus, no estimates can be made of the actual sulphide content of a gravity anomaly as opposed to the theoretical content calculated by integrating the anomaly curve. For many of us such a statement will come as a surprise, especially for a method that is reported to be quantitative. In actual fact, gravity can give an estimate of the sulphide content only if we have either previous and independent knowledge of the presence of sulphides or a probability function defining the presence of the sulphides under the anomaly (St-Amant and Gaucher, 1980, p. 1196). Of course, if the anomaly is big and the noise level evidently low, the likelihood of sulphides is high, but we do not know how high.

3. ESTIMATING THE SULPHIDE CONTENT FROM SEVERAL SURVEYS

3.1. Combining Magnetic and Electromagnetic Surveys

In the early days of electromagnetic surveying in Canada, magnetics were used extensively as a criterion for deciding whether a given electromagnetic conductor really existed. Often the early ground dip angle surveys could not discriminate between edges of swamps and bona fide bedrock conductors, especially if the survey was done at a single frequency. Airborne surveys also suffered from poor selectivity if two-frequency out-of-phase surveys were flown; sometimes even only a single frequency out of phase was flown. Even the airborne surveys·flown by Rio Tinto using both real and imaginary components were plagued by air turbulence noise on the real component. In those days, coincident magnetic peaks helped to screen out the overburden conductors. Today, except during VLF surveys, no one needs to rely on magnetics to confirm good bedrock conductors, whereas 'poor' bedrock conductors can be reliably distinguished from overburden by an IP profile.

A few companies still traditionally use magnetics to select the conductors most likely to bear economic mineralisation. A compilation of the Canadian orebodies by the writer has shown that there exists about the same proportion of 'non-magnetic' mines (30%) as there are non-magnetic conductors (50%). Indeed, the largest of the Canadian orebodies, Texas Gulf, is non-magnetic. Companies looking for nickel may be justified in selecting conductors on the basis of their magnetic response in the hope of selecting the nickeliferous mineralisations, but few companies are likely to refuse any orebody.

Figure 7 shows a compilation done from our bank of diamond drill holes to show the correlation between the magnetic anomalies over conductors and the content of sulphides. The procedure has been described in the graph showing width versus conductivity (Fig. 4). The compilation shows almost no correlation between the magnetic ano-

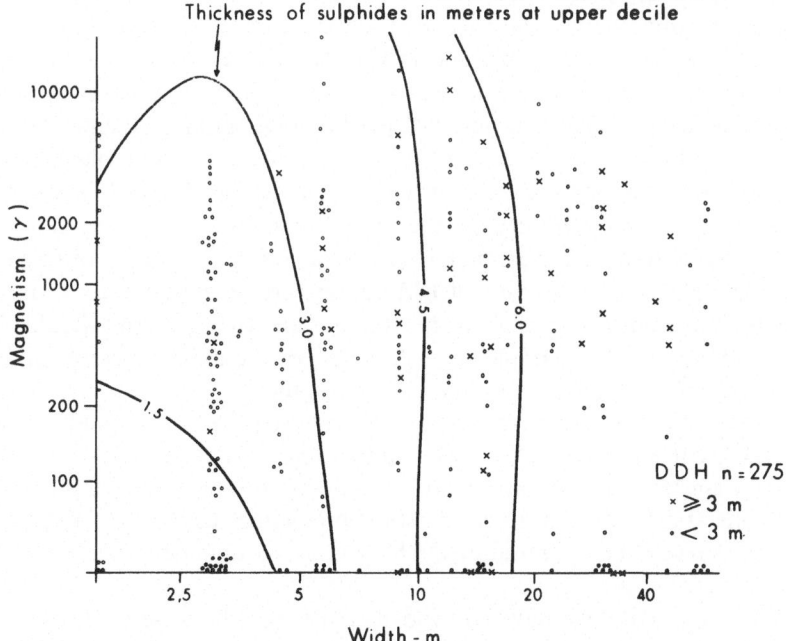

FIG. 7. Sulphide content of 275 diamond drill holes plotted according to the magnetism and width of the EM conductors tested. Sulphide content of 275 conductors tested by DDH, contoured at the upper decile as explained in Fig. 4. Magnetism is plotted along the vertical axis and width of the conductor along the horizontal axis. The diagram shows that magnetism does not seem to have a relationship to the sulphide content in the conductors, at least for conductors above the upper decile. The value of the contours was adjusted to correspond to Fig. 4. Approximate logarithmic scales.

malies of the conductors and their sulphide content, the contour lines being essentially vertical. The average content of sulphides increases slightly in conductors associated with anomalies between 1500 and 5000 γ, but if we consider only the better 10% of the conductors in any range of magnetism values, the advantage disappears.

For those companies which prefer selecting their conductors by magnetism, it may be a consolation that as only one conductor in three is non-magnetic, not much is lost by neglecting the non-magnetic ones. However, the biggest Canadian base metal orebody, Texas Gulf, should haunt some memories: it was an isolated, perfect target, 15 miles from the Timmins airport, but it was non-magnetic!

3.2. Combining Gravity and Electric Surveys

Since the mid-1950s, selecting EM conductors by gravity has been practised, at least in Sweden and in Canada. For example, the Mobrun deposit in the Noranda district was identified in 1954 as a likely massive sulphide-bearing DDH target before diamond drilling (Seigel *et al.*, 1957). Today a number of companies use gravity for that purpose, as witnessed by the popularity of gravity meters and hydrostatic electronic level instruments. Gravity is the only 'objective' screening method that is likely to detect most major massive sulphide orebodies among the swarms of tens of thousands of EM conductors. Any bona fide bedrock conductor identified either from the air or the ground should be checked by either a sample (a drill hole) or a gravity traverse; otherwise there was no purpose in finding the conductor.

Up to this point we have suggested how to predict sulphides from independent surveys: either EM conductors or IP anomalies, or from gravity surveys. In this section we shall describe how combining gravity with electrical surveys improves our prediction capability. Even so, gravity surveys have limitations, which must be understood in order to optimise their use.

To recapitulate, we have concluded that today, because of the lack of data banks, the quantity of sulphides cannot be predicted from a gravity survey, only how often a false anomaly of a certain size is likely to be encountered. On the other hand, at least for EM conductors in the Canadian shield, there are data banks which allow us to predict at the upper decile the quantity of sulphides likely to be present. By combining both approaches we can predict the sulphide content of a given combination of anomalies, describe why there are always fewer sulphides than expected, and propose how to deal with the remaining uncertainty in the decision tree of exploration.

The fundamental formula describing the expected sulphide content E of a conductor having a coincident gravity anomaly was devised by St-Amant and Gaucher (1980, p. 1192) as:

$$E(S_M/S_E, S_G) = S_G - \frac{\gamma^2}{S_E} + \gamma \sqrt{\frac{2}{\pi}} \frac{\exp[-(\gamma\sqrt{2}/2S_E - S_G/\gamma\sqrt{2})^2]}{1 - \mathrm{erf}(\gamma\sqrt{2}/2S_E - S_G/\gamma\sqrt{2})}$$

We shall refer the reader to the original article for the meaning of the symbols as the formula is complex; even so, already several simplifying assumptions have to be made in order to write it: for example, that the sulphide body is a vertical dike buried at a certain depth, etc. By fixing an additional number of parameters, such as the depth of overburden at 9 m, and by making a number of other reasonable assumptions, we can represent the expected sulphides graphically as in Fig. 8. We plot the observed gravity anomaly along the X-axis and the corresponding

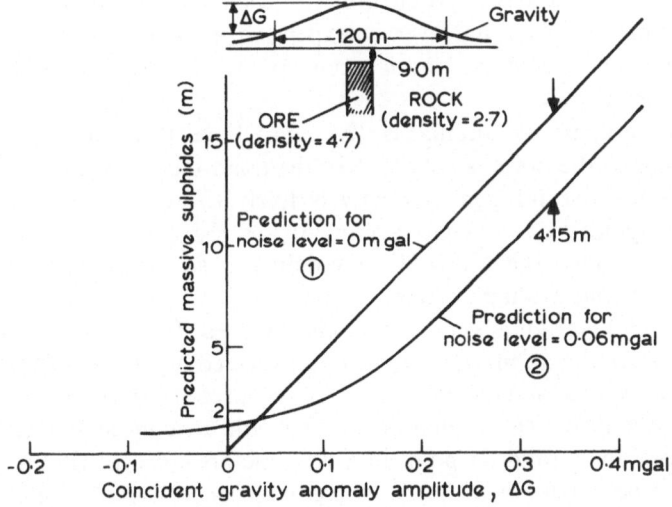

FIG. 8. Estimation of sulphides in a good conductor with a given gravity anomaly, with or without gravity noise. Curve 1 would predict sulphide content if there was no noise or if we knew that sulphides were present before a gravity survey was performed. Curve 2 predicts the sulphide content under a given gravity anomaly associated with a wide, excellent EM target. Such targets, if drilled indiscriminately, would contain at least 5 m of sulphides, one time out of ten. Curve 2 is drawn for the usual gravity noise level (0.06 mgal) observed in the Canadian shield. Curve 2 would diverge even further from curve 1 if the conductor was poorer (narrower) or if the gravity noise was higher. (Courtesy: Soquem.)

expected sulphides along the Y-axis. Two examples are illustrated:

1. Curve 1 represents how explorationists usually interpret information provided by gravity, that is, assuming no noise. They
assume that the amplitude of the gravity anomaly corresponds
directly to a given thickness of sulphides.
2. Curve 2 represents more realistic assumptions, expressed by our
equation where there is gravity noise and where gravity is used to
select among EM conductors for drilling. This line is valid only for
a given quality of EM conductors, here for a good conductor of
conductivity higher than 20 mho.

The equation expresses the following:

(a) A large gravity anomaly (over 0.2 mgal) correlates well with
sulphides. However, a 0.2 mgal anomaly is expected to contain an
average of only 4 m of sulphides instead of the 9 m that it would
contain if there was no noise. This deficiency for bigger anomalies
is constant and is proportional to the square of the gravity noise,
which implies that every inexpensive method should be used to
reduce the noise level, such as good levelling and terrain
corrections.
(b) Small gravity anomalies (0.1 mgal) have little diagnostic value
unless the noise is very low. In the noise measured in our surveys,
a positive 0.1 mgal anomaly is likely to contain twice as much
sulphides as a negative anomaly of the same value. For such
anomalies, the width of the conductor is a much better guide to
selecting drilling targets.
(c) The equation is general and can be used if there is no conductor
under the anomaly, but then an expected sulphide content has to
be defined independently. We have already suggested a sulphide
content for IP anomalies, and it would be conceivable to assume
one for a random point in a volcanic area, except that it is likely
to be much lower than over a conductor. *Ipso facto* the equations
then describe a simple gravity survey, where we have already
concluded that a much stronger anomaly of 0.4 or 0.5 mgal was
necessary before deciding to drill.

To be able to use conveniently the equation of expected sulphide
content of a gravity anomaly associated with a conductor of a given
sulphide content, we have prepared a graphical solution which should be
valid for a broad number of reasonable assumptions, the most critical of

which is the regional noise level, assumed to be 0.06 mgal as measured in our compilations. To use the graphical solution illustrated in Fig. 9, one must first measure the amplitude of the gravity anomaly from a cord of 120 or 125 m, then estimate the sulphide content of the target at the upper decile. If the conductor selected has a width of 10 m and a conductivity of 100 mho, Fig. 4 predicts a sulphide content at the upper decile of somewhat over 5 m. We plot the gravity anomaly of 0.45 mgal in column A, the expected sulphide content at the upper decile in column C; we can then draw a straight line between the two points and read off in column B the average sulphides that will be intersected, somewhat over 5 m thick. Thus, such a gravity high raises the sulphide content of the

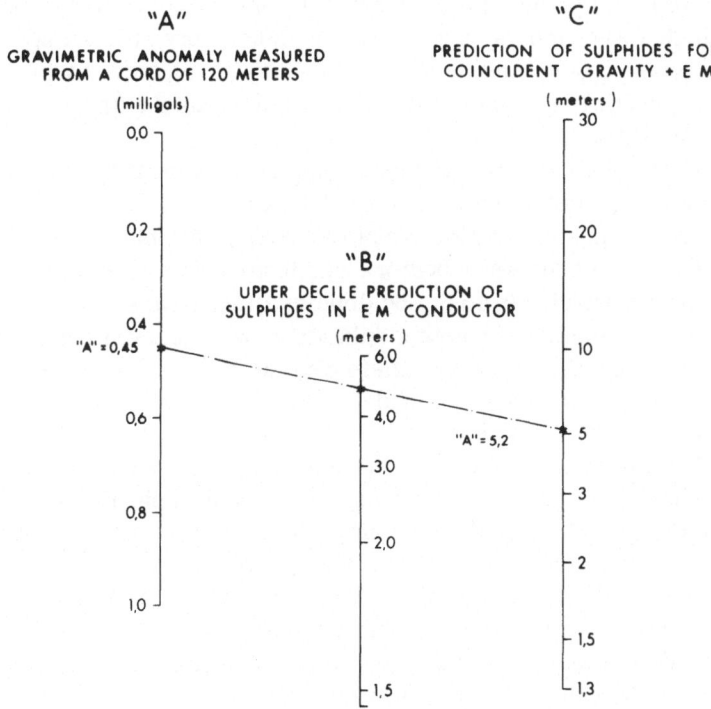

FIG. 9. Nomogram for estimation of sulphide content of an EM or IP target from its width and its gravity anomaly. A good EM anomaly with a coinciding gravity peak of 0.45 mgal would contain an average of 5 m of sulphides; however, in the absence of gravity information it would contain as much sulphides only one time out of ten.

conductor to an average of 5 m, rather than just intersecting 5 m, one time out of ten. The average sulphide content of such a conductor with a gravity high is likely to be ten times higher than if we did not know its gravity response. The increase may be even more dramatic for the narrower conductors.

4. CONCLUSIONS

In this study we have reviewed in detail the methods of estimating the sulphide content of diamond drill targets, the potential orebodies of any mining exploration venture. We shall now restate the main conclusions of the study:

(a) Even if all predictions as to sulphide content are characterised by high dispersion, it is possible to reduce very substantially the average cost of finding a given quantity of sulphides by properly engineering the exploration approach towards the upper decile of the targets.

(b) Width of conductors and of IP anomalies is probably an important prediction criterion of the sulphide content. Conductivity and magnetism are much less diagnostic, even if they have been used to discriminate between bedrock conductors and conductive overburden, which can today be done by other means.

(c) Gravity anomalies whose amplitude is two and, even better, three times higher than background noise can be an efficient prediction criterion; but smaller anomalies, within the noise background, are likely to be misleading.

(d) The proper allocation of the total budget between finding and selecting targets versus testing them by drilling, is for the exploration manager the most important and the easiest parameter to optimise and to supervise, and yet the most frequently overlooked. Without exception, all our simulations and studies conclude that at least half the budget should be spent on testing the targets selected by surface estimations, and that it is impossible to 'save' by drilling or trenching less than that.

(e) Finding mines, and even finding sulphides, is still somewhat of an art, and great liberty should be given to geologists to use their imagination and their hunches so as to integrate past knowledge of economic mineralisations into the exploration venture, giving proper weight, for example, to known showings, floats of ore found by prospectors or to results of past drilling campaigns. The only budget that managers should not allow to be reduced or preempted is the drilling budget. Drilling funds should be used promptly to test whatever targets are defined, even during the project, as such interim drilling results provide useful information which can guide the venture. If one waits to the end, then budgets may be cut and targets are left untested.

The statistical treatment of exploration data in predicting the sulphide content of potential orebodies is a new and controversial subject. We fully realise that much more work remains to be done before a well-established body of knowledge is amassed; and as in any new area, new rules will still have to be found. We hope that this study will be found useful and that it will encourage others to compile their field results and publish their conclusions. For those contemplating such an endeavour, we suggest optimising not for the 'average' but for the upper decile—for the exceptional target. Only by doing so did we begin to perceive some order in our exploration data banks.

ACKNOWLEDGEMENTS

We wish to thank the following group of companies which provided material for this paper and made these compilations possible: Soquem (Quebec Mining Exploration Company), where the main body of the study was done; but also Pennaroya, Cominco, New Jersey Zinc and the Geological Survey of Canada. We also wish to thank the many individuals who formally or informally collaborated in many phases of this study. In particular, we should like to mention C. Carbonneau, G. Tikkanen, H. V. McMurry, H. Meyer, J. Hamilton, J. Lajoie, A. Nadeau, R. Lambert, P. de Bretizel, M. St-Amant, D. C. Gagnon, G. Gélinas, R. St-Pierre, D. Brulotte, R. Assad and A. Poitras.

REFERENCES

BOLDY, J. (1979) Exploration discoveries, Noranda District, Quebec (case history of a mining camp). In *Geophysics and Geochemistry in the Search for Metallic Ores*, Geol. Surv. Can., Econ. Geol. Rep. 31, pp. 593–602.
BROCK, J. S. (1973) Geophysical exploration leading to the discovery of the Faro deposit. *Can. Mining Metall. Bull.* **66**(38) 97–116.

CRONE, D. (1979) Exploration for massive sulphides in desert areas using the ground pulse electromagnetic method. In *Geophysics and Geochemistry in the Search for Metallic Ores*, Geol. Surv. Can., Econ. Geol. Rep. 31, pp. 745–55.

GAUCHER, E. (1980) Levé de polarisation provoquée, gisement de Phelps Dodge, Canton La Gauchetière. Rapport interne, pour Exploration Noranda Ltée.

GAUCHER, E. (1981) La fièvre de l'or, en profitons-nous? *Bull. Géol. Québec* No. 43, 62–6.

GAUCHER, E. and NADEAU, A. (1972) Echantillonnage des travaux d'exploration minière de 200 cantons de l'Abitibi Nord. Commission Géologique du Canada, Etude 72–13.

GAUCHER, E. *et al.* (1972) Optimization of exploration expenditures. Internal report at Soquem, Cominco, New Jersey Zinc and Pennaroya.

HOOD, P. (1980) Mineral exploration trends and developments in 1979. *Can. Mining. J.* **101**(1), 20–63.

KOULOMZINE, T. and BROSSARD, L. (1957) Magnetometer surveys in the area of the Bourlamaque batholith and its satellites. In *Methods and Case Histories in Mining Geophysics: 6th Commonwealth Mining and Metallurgical Congress*, Mercury Press Co., Montreal, Quebec.

LAJOIE, J. and KLEIN J. (1979) Geophysical exploration at the Pine Point Mines Ltd. zinc–lead property, Northwest Territories, Canada, In *Geophysics and Geochemistry in the Search for Metallic Ores*, Geol. Surv. Can., Econ. Geol. Rep. 31, pp. 653–64.

PARASNIS, D. S. (1975) *Mining Geophysics*, Elsevier, Amsterdam.

PATERSON, N. (1979) Geophysical prospecting for uranium in the Athabasca Basin, Preprint, personnal communication.

REED, L. E. (1979) The discovery and definition of the Lessard base metal deposit, Quebec. In *Geophysics and Geochemistry in the Search for Metallic Ores*, Geol. Surv. Can., Econ. Geol. Rep. 31, pp. 631–9.

REED, L. E. (1981) The airborne electromagnetic discovery of the Detour zinc–copper–silver deposits, northwestern Quebec, *Geophysics* **46**(9), 1278–90.

ST-AMANT, M. and GAUCHER, E. (1980) Probabilistic evaluation model of an electromagnetic target with a given gravity anomaly. *Geophysics* **45**(7), 1184–96.

SCOTT, J. S. (1948) *Quemont Mine, Structural Geology of Canadian Ore Deposits*, Canadian Institute of Mining and Metallurgy, Mercury Press, Montreal, Quebec.

SEIGEL, H. O. *et al.* (1957) Discovery of the Mobrun Copper Ltd. sulphide deposit, Noranda mining district, Quebec. In *Methods and Case Histories in Mining Geophysics: 6th Commonwealth Mining and Metallurgical Congress*, Mercury Press Co., Montreal, Quebec.

SNYDER, D. D. and MERKEL, R. H. (1977) Induced polarization measurements in and around boreholes. In *Induced Polarization for Exploration Geologists and Geophysicists*, Department of Geosciences, University of Arizona, Tucson, pp. 161–220.

SPITERI, G. and BARIA, O. R. (1981) Geological interpretation of the Detour Lake gold deposit. Presented at the 1981 Prospectors Convention, Toronto, Ontario.

SUMNER, J. S. (1979) The induced polarization method. In *Geophysics and*

Geochemistry in the Search for Metallic Ores, Geol. Surv. Can., Econ. Geol. Rep. 31, pp. 123–33.

WARD, S. H. (1979) Ground electromagnetic methods and base metals. In *Geophysics and Geochemistry in the Search for Metallic Ores*, Geol. Surv. Can., Econ. Geol. Rep. 31, pp. 45–62.

Chapter 2

A SWEEP-FREQUENCY ELECTROMAGNETIC
EXPLORATION METHOD

I. J. WON

North Carolina State University, Raleigh, North Carolina, USA

SUMMARY

A prototype sweep-frequency electromagnetic system has been developed and field-tested. The system, housed in a moving ground vehicle, operates in transit with a loop–loop transmitter and receiver configuration. The transmitter produces logarithmically sweeping harmonic waves between 500 Hz and 100 kHz lasting a few seconds at a maximum output of about 2 kW. Secondary field amplitude and phase spectra are measured, digitised and stacked several times to enhance the signal-to-noise ratio before being transcribed on a disk storage. The entire operational cycle is automated and controlled by an on-board microcomputer. The resultant data are displayed as continuous spectral profiles as a function of distance which may be interpreted as a cross-sectional conductivity structure of the earth in the survey area. Theoretical studies involving several simple models, including a conductive circular cylinder in a conductive half-space, and a horizontally layered earth, suggest that there exist many intuitive correlations between a given geological model and its spectral profile. Laboratory-scale model experiments using a sweep-frequency source also support theoretical predictions of the spectral behaviours.

1. INTRODUCTION

The electromagnetic (EM) exploration system employing a single frequency or a few discrete frequencies up to the RF band has long been in use in the mining industry for detecting mineral deposits which are usually anomalous in electrical conductivity. The method, often referred to as the induction EM method in mining geophysics, involves the propagation of a time-varying low-frequency EM field in and over the earth.

Ideally, a geophysical exploration survey not only locates a sought object but also delineates the three-dimensional (3-D) geological structure which surrounds the object. Thus, it is desirable to develop an EM method which can rapidly and economically map the entire 3-D subsurface variations. These variations may be due to mineral deposits, ground water, stratigraphic or tectonic structures, and any other associated geological conditions having some contrast in electrical properties.

One obvious approach which can provide both deep penetration and high vertical resolution is an EM method employing a wide-band sweep-frequency source. The depth of penetration is mainly determined by the source frequency and ground conductivity. The relationship is shown by the nomogram in Fig. 1. Therefore, applying a sweep-frequency EM field is equivalent to a depth sounding.

The idea of using a wide-band multi-frequency source in the EM method is not new either theoretically or experimentally. To mention a few recently reported field experiments, Ryu et al. (1972) used 14 discrete frequencies between 200 Hz and 10 kHz and measured tilt angle, ellipticity and the modulus of wave tilt to explore for ground water in the Santa Clara Valley, California. Similarly, Ward et al. (1974, 1977) used 14 discrete frequencies between 10.5 Hz and 86 kHz to explore a sulphide mineral deposit in Cavendish, Ontario.

One of the drawbacks of these systems is that each frequency requires a separate operation. Thus, without introducing fast frequency-multiplexing and power-switching schemes, such a system cannot be used on a moving vehicle. This eliminates its use as a fast (possibly airborne) reconnaissance tool.

There are two main obstacles which have impeded the development of a continuous-frequency EM system: (1) poor theoretical understanding of wide-band induction phenomena, and (2) lack of instrumentation for a high-powered wide-band system with associated field data processing. The theoretical understanding of the wide-band diffraction problem as

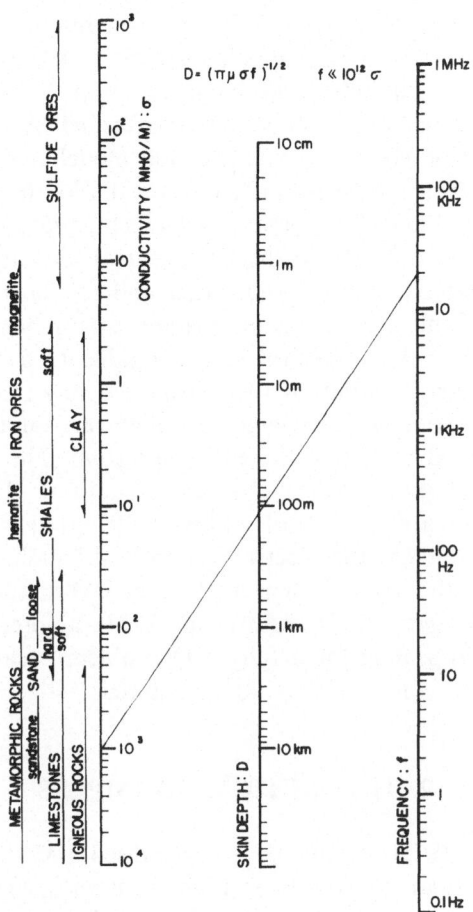

FIG. 1. Relationship between source frequency, ground conductivity and depth of penetration. Magnetic permeability, μ, is assumed to be that of the free space. For example, if the source frequency sweeps from 100 Hz to 100 kHz in a typical igneous rock area, the depth of exploration (skin depth) ranges from about 40 to 1500 m. (From Won, 1980. Courtesy: Society of Exploration Geophysics.)

applied to EM exploration, although steadily improving, is still primitive except for a few simple geometries. The available theories often become intractable, particularly when all media involved must be considered finitely conductive. On the other hand, the instrumentation difficulties now appear to be quite surmountable with present electronic technology. The large amount of data to be collected in a sweep-frequency EM system is no longer a concern with current high-volume data-logging

techniques, and is rather desirable since a maximum amount of information is gained without appreciable increase in field efforts.

The wide-band diffraction problem essential for developing any
continuous-frequency EM system can be attacked by two approaches,
namely, solution of Maxwell's equation and model study. Although the
method is based on the well-founded classical EM theory (Wait, 1962;
Grant and West, 1965; Keller and Frischknecht, 1966; Ward, 1967; and
others), the application of the theory to a realistic earth with a finite
conductivity contrast between the target and the host medium is extremely difficult, and available solutions are usually band-limited.

A rigorous theoretical spectral study suitable to a sweep-frequency
EM system requires a solution to be valid for any frequency with no
constraints on the EM parameters, so that the solution can be re-
petitively evaluated to obtain the entire spectrum. The numerical techniques developed most recently by several authors (Hohmann, 1971,
1975; Parry and Ward, 1971; Won and Kuo, 1975; Lajoie et al., 1975) are
powerful in the study of the spectral response, although computationally
expensive. While the finite-element method for EM diffraction appears to
be promising (Coggon, 1971; Rijo et al., 1977), at present the method
does not seem appealing for a fully 3-D problem because of the large
computer storage, time and resolution required.

2. THEORETICAL EXAMPLES

Won and Kuo (1975) formulated a generalised EM diffraction theory
involving more than two media (e.g. air, host rock, target body), each
having an arbitrary electrical conductivity, a dielectric permittivity and a
magnetic susceptibility. Won (1980) discussed the case of a circular
cylinder of an infinite length in a conductive half-space, as reproduced in
Fig. 2. The geometry shown in the inset depicts a circular cylinder having
a radius of 50 m and a conductivity of 100 mho/m buried in a half-space
of a conductivity of 0.01 mho/m. The source is an insulated wire carrying
1 A placed on the ground directly above the cylinder while the scattered
fields are measured at an altitude of 50 m. The figure shows the peak
spectral responses of various secondary fields computed at seven different
frequencies between 10 Hz and 30 kHz for three different depths (20, 60
and 100 m) of the cylinder.

While the residual magnetic field rapidly decreases as frequency
increases, the electric field peaks at a particular frequency which evi-

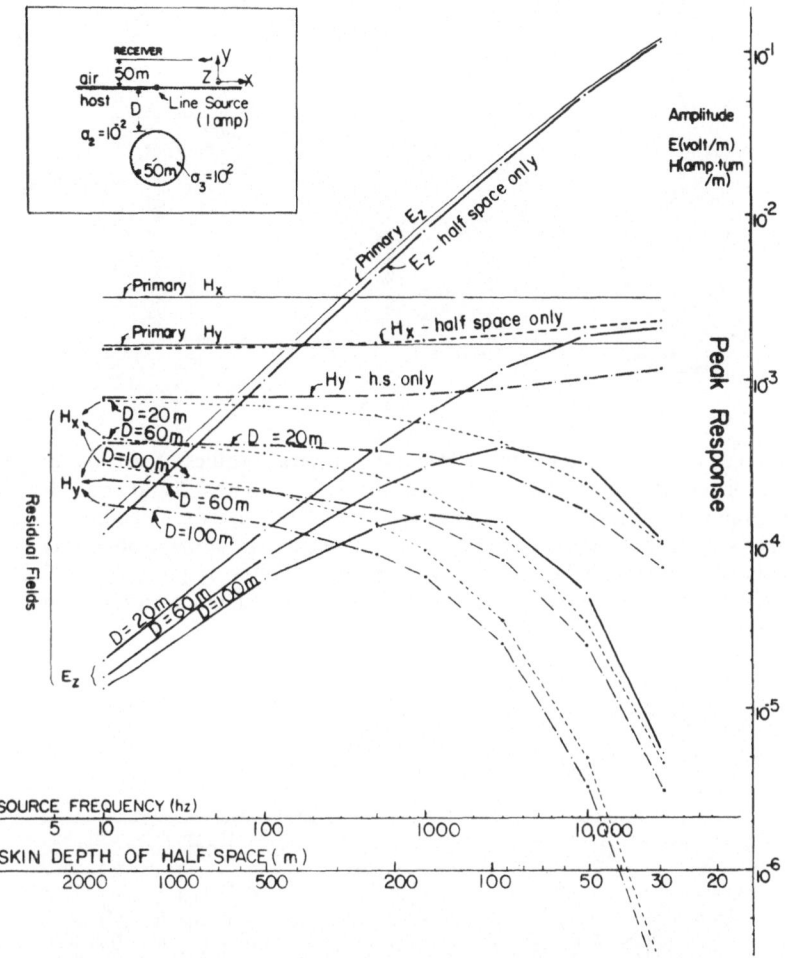

FIG. 2. Peak amplitude response as a function of frequency for a circular cylinder in a half-space at three different depths, 20, 60 and 100 m. Skin depth of the half-space is shown along the frequency axis. (Source: as Fig. 1.)

dently depends on the depth of the cylinder. Since the magnetic field is measured using a loop antenna, the induced EMF which is proportional to frequency will also possess similar peaks. The peaks occur at the frequency at which the corresponding skin depth of the half-space is slightly greater than the actual depth of the target. This can be theoretically explained by considering a constructive interference of a dissipative EM plane wave, as discussed in detail by Won (1980).

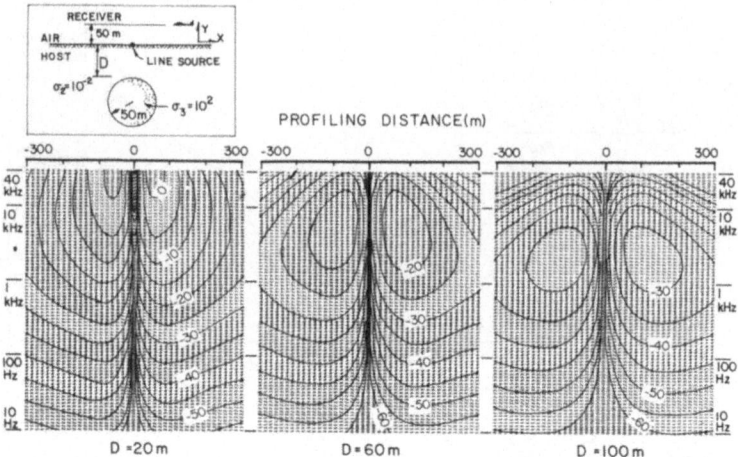

FIG. 3. Theoretically computed amplitude spectral profiles of vertical magnetic field for the situation shown in the inset for various depths, 20, 60 and 100 m. The horizontal axis denotes the profiling distance of the receiver extending 300 m on either side of the circular cylinder located at the centre of the profile. Any single-frequency profile may be obtained by selecting a specific frequency on the vertical axis. The contour interval is 5 dB with an arbitrary reference. (Source: as Fig. 1.)

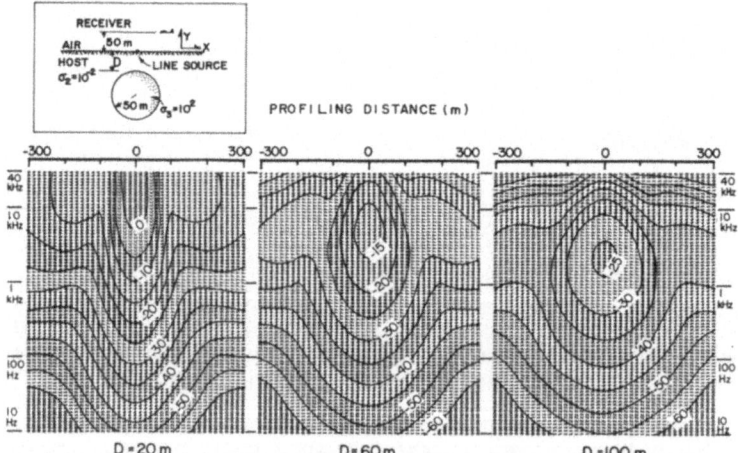

FIG. 4. Theoretically computed amplitude spectral profiles of horizontal magnetic field for the situation shown in the inset for various depths, 20, 60 and 100 m. The horizontal axis denotes the profiling distance of the receiver extending 300 m on either side of the circular cylinder located at the centre of the profile. Any single-frequency profile may be obtained by selecting a specific frequency on the vertical axis. The contour interval is 5 dB with an arbitrary reference. (Source: as Fig. 1.)

As a method of presenting the secondary field data, Won (1980) compiled the profiles for many different frequencies and reduced them to a single spectral cross-section. Figures 3 and 4 show such spectral cross-sections of the vertical and horizontal magnetic fields due to a cylinder at three different depths, 20, 60 and 100 m. Such spectral maps, which are somewhat similar to the geological cross-section derived from the reflection seismic exploration method, may be a very useful means of displaying and investigating a large amount of data at a glance.

3. RESULTS FROM SCALED MODEL EXPERIMENTS

In order to test the theoretical results, Won (1980) experimented with a scaled laboratory sweep-frequency EM system which is shown schematically in Fig. 5. The conductive earth was simulated with saline water, while the target was simulated with a 1.5 cm thick graphite slab 110 × 32 cm.

A carriage on which a transmitter and a receiver were mounted was electrically driven along a fibreglass I-beam. Measurements were made at 2 cm intervals across the target. The transmitter signal consisted of a continuous logarithmic sweep from 4 kHz to 4 MHz for a duration of 30 s. The final amplitude and phase spectra were recorded on an X–Y plotter in real time (and digitised for later machine contouring). The frequency axis of the plot was provided by a triangular ramp voltage which was used to generate the sweep source through a voltage-to-

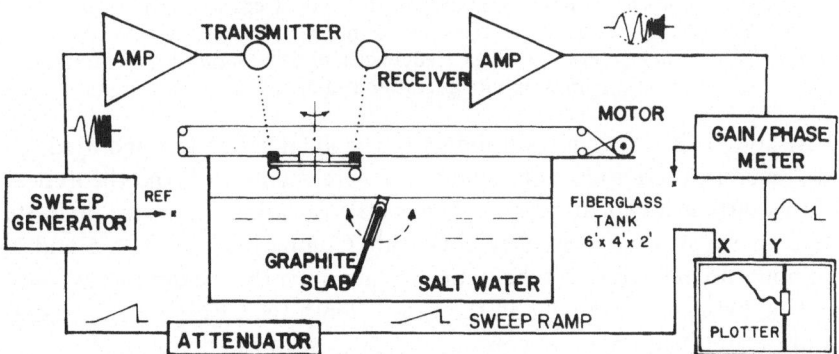

FIG. 5. Block diagram of the EM model system employing a wide-band sweep signal. The field amplitude may be measured in either absolute value or in percentage anomaly compared with the primary field. (Source: as Fig. 1.)

frequency converter. The effect of the tank edge was found to be negligible with the profile distance.

Figures 6 and 7 show the experimentally obtained spectral responses for a vertical graphite slab. The half-space response along with other ambient responses was determined by placing the sensor assembly far from the graphite slab along the profile path. The amplitude and phase spectra at this position were subtracted afterwards. Therefore, these spectra represent only those of residual signals due to the presence of the graphite slab.

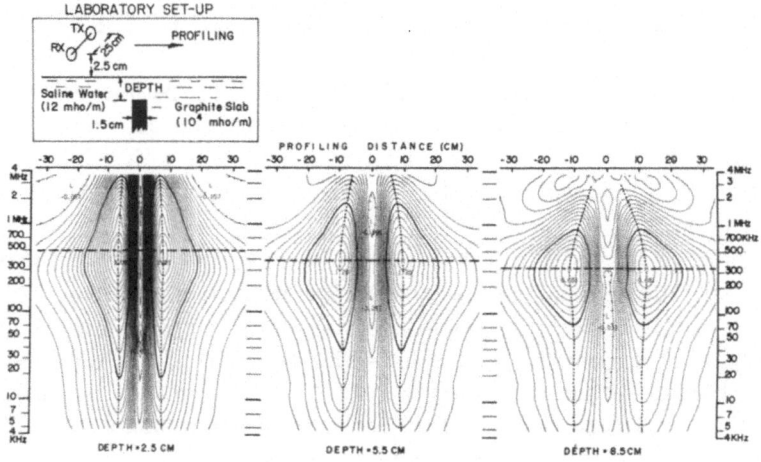

FIG. 6. Experimentally obtained amplitude spectra of magnetic field of a buried dike in a conductive half-space employing a vertical coils system in a scaled model. Although the set-up is not the same as in Fig. 3, general agreements in the spectral behaviour are evident. Peak frequencies and half-anomaly contours are shown in heavy lines as interpretational aids. (Source: as Fig. 1.)

Although the experimental model is not identical to the one used for the theoretical computation, qualitative agreements with the theoretical results are evident in terms of spectral variation in space and its dependence as a function of target depth. Comparing Figs 3, 4, 6 and 7, we note the following: (1) the frequency at which the maximum response occurs decreases gradually as the target depth increases; (2) the spatial width of the half-anomaly contour increases with the depth; and (3) the spectral sections of Figs 4 and 7 for the horizontal magnetic field somewhat resemble the geometrical configuration of the target in the subsurface.

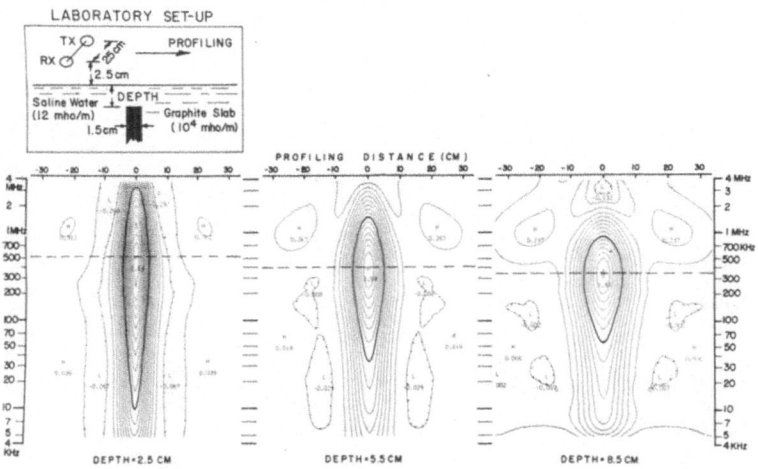

FIG. 7. Experimentally obtained amplitude spectra of magnetic field of a buried dike in a conductive half-space employing a horizontal coils system in a scaled model set-up. (Source: as Fig. 1.)

The spectral characteristics as described above have been qualitatively known and are physically understandable. Rather, the main advantage of such wide-band spectral data is the method of cross-sectionally displaying the entire data in order to provide some intuitive interpretations.

While EM techniques have been in use mostly for minerals exploration, the method in general has a potential application for structural and stratigraphic mapping purposes as well. The latter application, if successful, would have wide usages in many geological problems, particularly if the results show cross-sectional earth structures in a visual format similar to those from the reflection seismic method.

The theoretical EM response of a layered earth has been studied by several authors. Frischknecht (1967) published a comprehensive article on the EM response of a two-layered earth in the presence of an oscillating magnetic dipole. The theory disregards the presence of displacement currents. Anderson (1974) extended it to a multi-layered earth model based upon the theory of Frischknecht, and developed computer programs for both the forward calculation algorithm for frequency response and the inversion algorithm based upon the Marquardt nonlinear least-squares technique.

Figure 8 shows the theoretical amplitude and phase profiles for the depicted geological model computed from Anderson's forward-

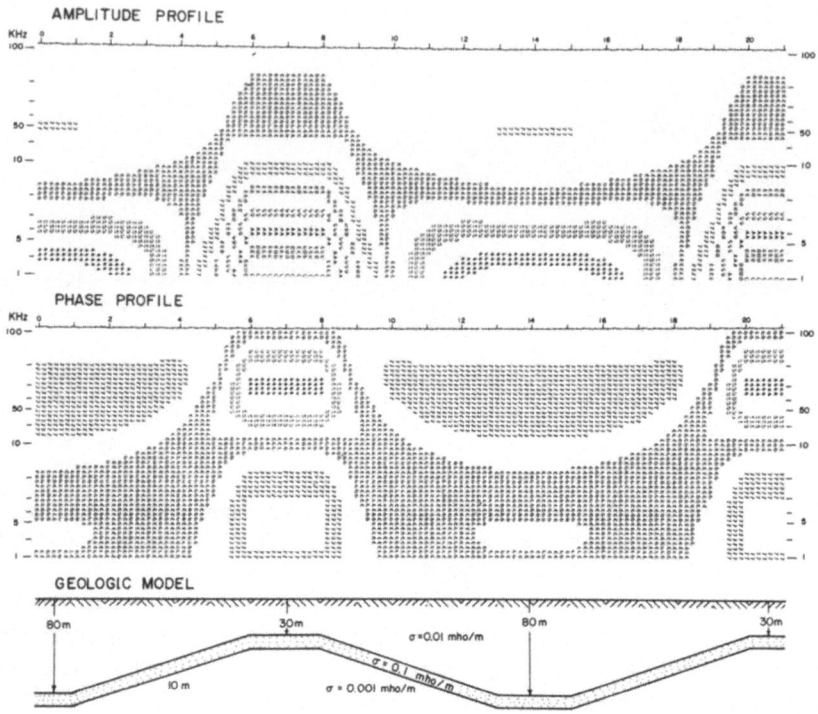

FIG. 8. Theoretical amplitude and phase spectra for the geological model shown,
assuming that below each station the earth is horizontally stratified.

calculation algorithm. The computed spectra are approximate in the
sense that they are calculated based upon the assumption that the layers
were horizontally stratified below each calculation point. The conductive
second layer has a constant thickness of 10 m. At each point the
amplitude and phase spectra are computed for 12 frequencies for each
decade between 1 kHz and 100 kHz. The transmitter and receiver are
assumed to be 10 m apart and of horizontal coplanar configuration at an
elevation of 2 m above ground.

The amplitude spectrum plotted here is derived as follows: after
computing the entire spectra for all stations, we first determined an
average spectrum by algebraically averaging the entire spectra; this
average spectrum was then subtracted from each original spectrum to
produce the 'residual' spectral profiles which are shown in this figure.
This process removes most of the half-space response, leaving only

'relatively' anomalous spectral features. (This process will be discussed again later when we encounter the field data.) These theoretical spectral profiles suggest that there exist intuitive relationships between the geological model and the resultant frequency–distance section, somewhat analogous to that of the reflection seismic method. Obviously, the nature of the correlation is complicated and needs further development.

One of the obvious advantages of the spectral profiling technique is to be able to classify any significant anomalies in terms of their subsurface distributions. Physically, an anomaly caused by a large and deep structure will be a broad feature in space and will appear towards the lower end of the spectrum, while an anomaly caused by a small and shallow structure will be a narrow feature in space and will appear towards the higher end of the spectrum. Transition of the spectral peaks as a function of the profile distance can enhance the subsurface features, discriminating against the effects of lateral variations.

4. A PROTOTYPE FIELD SYSTEM

Encouraged by the limited theoretical and laboratory experimental results, we proceeded to build a prototype field system. Figure 9 shows the block diagram of the system. The system may be functionally divided

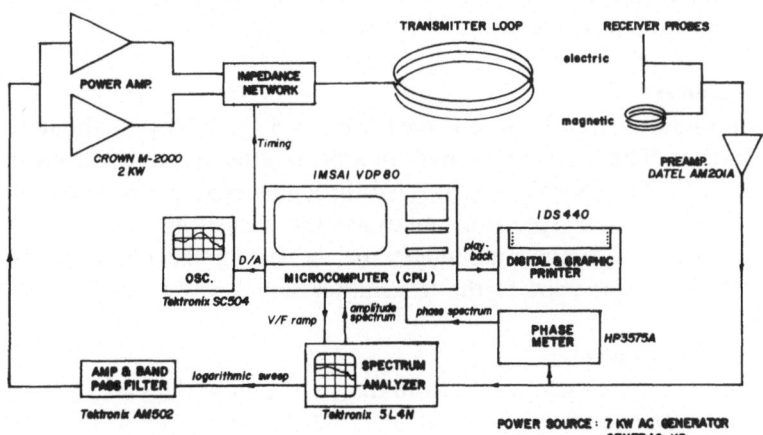

FIG. 9. Block diagram of the sweep-frequency electromagnetic system.

into five subunits: (a) transmitter, (b) receiver, (c) system control, (d) data acquisition and (e) playback. A brief description of each subunit follows.

4.1. Transmitter

A sweep-signal generator built into the spectrum analyser produces a constant-amplitude sweep whose frequency is logarithmically swept within a predetermined frequency range. The unit is driven externally by a microcomputer which generates a linear voltage ramp. This ramp is used by a voltage-to-frequency converter for generating the sweep and provides the frequency axis during measurements. The sweep output of the spectrum analyser is fed into a power amplifier whose output is connected to a radiating transmitter loop. The loop consists of about 20 turns of seven-strand No. 14 copper wire with a thick (THD) insulation, wound on a 1.8×3.0 m aluminium frame. The loop is mounted on a small trailer which houses the two power amplifiers and an electrical power generator. The loop can be positioned either horizontally or vertically. Other than the power amplifiers and the generator, the entire system is housed in a small truck.

The loop carries a total resistance of about $1.1\,\Omega$ and a total inductance of about 5.5 mH. Since the output impedance of the main power amplifier is about $4\,\Omega$, the lowest frequency the amplifier can accommodate is about 500 Hz with the present loop. This limit can be lowered if we increase the number of turns. However, an increase in turns will result in a power decrease towards the high-frequency end of the spectrum. Presently, the system can be operated from 500 Hz to 100 kHz with a maximum power of 2 kW at 500 Hz.

4.2. Receiver

The secondary magnetic field is measured by a small loop of about 30 cm in diameter. The loop is mounted on a beam attached to the front hood of the truck. The received signal is fed through a modular wide-band pre-amplifier into the spectrum analyser for measuring the amplitude spectrum and into the phasemeter for measuring the phase spectrum. The phase is referenced to the transmitter signal waveform.

4.3. System Control

The entire measurement cycle, from the activation of the sweep generator to the data transcription, is controlled by the microprocessor unit. The measurements can be repeated either manually at each measurement site

or automatically at a preset time interval for a moving vehicle operation.

4.4 Data Acquisition

Figure 10 shows the timing sequence for the system operation. At each
measurement point a prescribed number of sweeps are generated and the
ground spectral responses in both amplitude and phase are measured. The
resultant data are arithmetically stacked in order to enhance the signal-
to-noise ratio. Since the resolution of the A/D converter is 12 bits, a
stack of, say, ten gives a resolution of 0.0024%. In reality, however, the
natural and cultural noise may degrade the signal quality considerably.

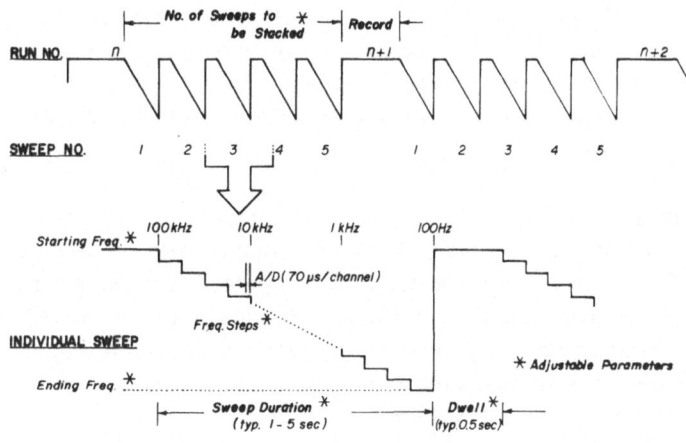

FIG. 10. Timing and sweep ramp configuration.

Each sweep consists of a number of stepwise changes in frequency,
typically between 50 and 500 steps. The stepping is accomplished by the
microcomputer which generates a linear algebraic series and, through a
D/A converter, feeds a stepping ramp voltage into the spectrum analyser.
The spectrum analyser, in turn, converts the ramp voltage into a
logarithmically sweeping sinusoidal waveform.

At the end of each step the received amplitude and phase spectra are
digitised and recorded on a temporary storage memory. By waiting until
the end of a step, we allow the system to stabilise sufficiently at the new
output frequency. As an example, for a sweep duration of 2 s employing

200 steps, each step lasts 10 ms. The signals are sampled and recorded during the last 0.07 ms. Each digitised spectrum is arithmetically added to the previous spectrum. When all sweeps are completed, the spectrum is then transcribed permanently on a disk storage.

In order to maintain a smooth power consumption cycle, the sweep is made from a high frequency to a low frequency. The output remains at the high-frequency end while idling between sweeps. With the present set-up we can achieve a minimum sweep duration of about 1.5 s. This limit is imposed by the time constant of the spectrum analyser. With an improved spectrum analyser it is believed that the entire sweep speed can be reduced to 1 s or less for an ultimate airborne operation. On the other hand, the lower limit of frequency also dictates the sweep speed: the lower the frequency, the longer the sweep time required.

The several timing parameters shown in Fig. 10 are variable and can be changed, particularly (1) the number of sweeps to be stacked, (2) the number of frequency steps within a sweep, (3) the sweep duration, (4) the starting frequency, (5) the ending frequency and (6) the dwell time between sweeps.

4.5. Playback
The field data can be played back immediately after a profile is completed. First, an average spectrum is computed from the entire profile. The average spectrum thus obtained contains both the overall system response and the mean geological response of the surveyed area. This average spectrum is then subtracted from each original spectrum, resulting in an 'anomalous' spectrum.

5. PRELIMINARY TEST RESULTS

Figure 11 shows a test profile made across two known diabase dikes in Triassic clay in Chatham County, North Carolina. Based upon shallow drill data, both dikes are believed to be almost vertical ($\sim 85°$ dip to the east, i.e. to the right of the profile). Other than these two dikes, the area is composed of various types of soils, mostly clay.

The vertical frequency axis may be converted into a pseudo-depth scale using the skin-depth relationship. If we take a clay conductivity of, say, 0.1 mho/m, the vertical scale would extend to about 90 m.

The contour map is a composite of 32 spectra spaced at intervals of 10 m, and employing a maximum-coupled horizontal coplanar coils

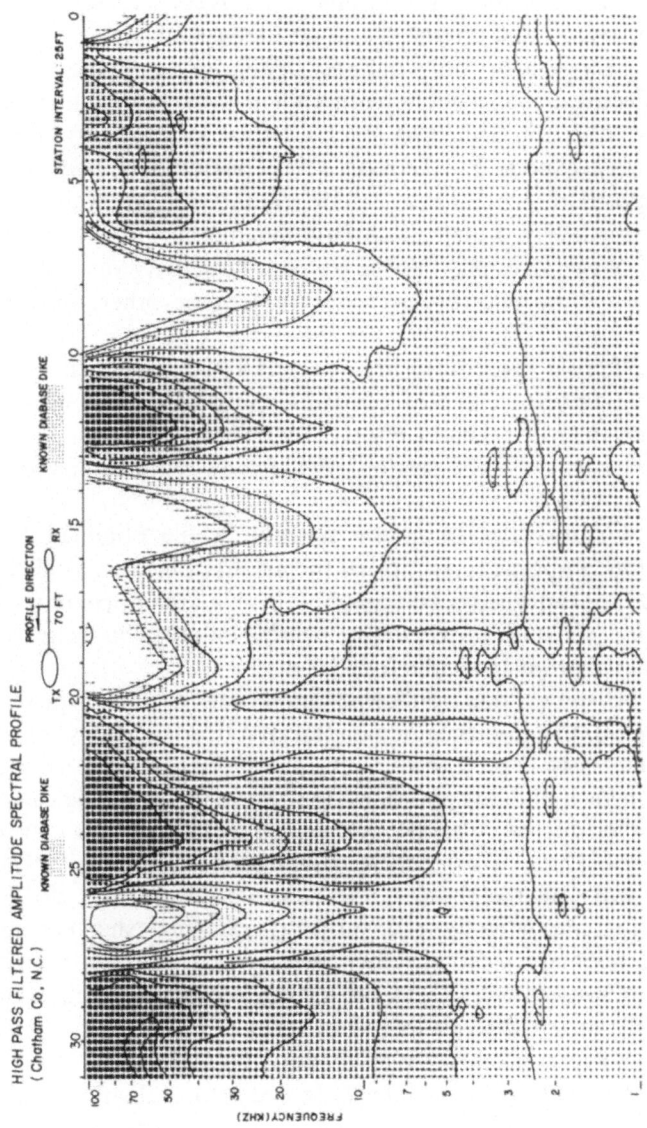

FIG. 11. Amplitude spectral profile over two known diabase dikes in a Triassic basin of North Carolina. From drill data, the dikes are believed to be dipping about 85° to the east (to the right). The dikes are hosted by various clay formations.

configuration. The plotted amplitude scale is in dB with a maximum anomaly of about 12% in the vicinity of the dikes with respect to the primary field. It is evident that the profile contains considerably anomalous spectral features on and around the target diabase dikes.

However, the profile also contains significant anomalies elsewhere. Since the electrical properties of neither the host clay nor the diabase are well known, it is not clear how much contrast in conductivity between them we may expect. Furthermore, clay possesses a wide range of conductivity (0.01 to ~1 mho/m) depending on its water content. Therefore, it is believed that the spectral fluctuations other than those in the vicinity of the dikes may be due to the conductivity variations within the clay formation, affected by water content or other mineralogical variations.

Various filtering procedures may be performed during the playback. Figure 12 shows a spectral profile containing only the shallow anomalies derived from the same data of Fig. 11. By removing all anomalies at the low-frequency end (deepest section), the shallow features are somewhat emphasised.

Figure 13 shows the amplitude spectral profile obtained across a known vein of graphitic schist hosted by metamorphic rocks chiefly composed of schist and gneiss. The graphite in this area appears to be disseminated on the surface. However, nearby outcrops indicate that the vein width at depth may be about 2–3 ft.

Figure 14 shows one unfiltered amplitude profile approximately 3500 ft over a metamorphic region locally overlain by sediments of variable thickness. The known cultural objects as well as available surface geological data are shown on the approximate geological section. The anomalies due to cultural objects such as bridges and culverts are easily recognisable because their high-frequency anomalies are big enough to wipe out the entire spectral section. In contrast, the geological anomalies are laterally and vertically extended showing spatially distributed anomaly patterns. The profile was made while driving at about 10 km/h speed by stacking five sweeps per station, each sweep lasting 2 s and having 70 discrete frequency steps.

The ground response may be affected by the orientation of the profile as well as the heading of the vehicle. In order to evaluate the effect, we obtained two-way profiles along a portion of the same road. Figure 15 shows the southbound profile while Fig. 16 shows the northbound profile. If the system reproduces correctly, the two profiles should be mirror images of each other. This appears to be roughly the case. Since

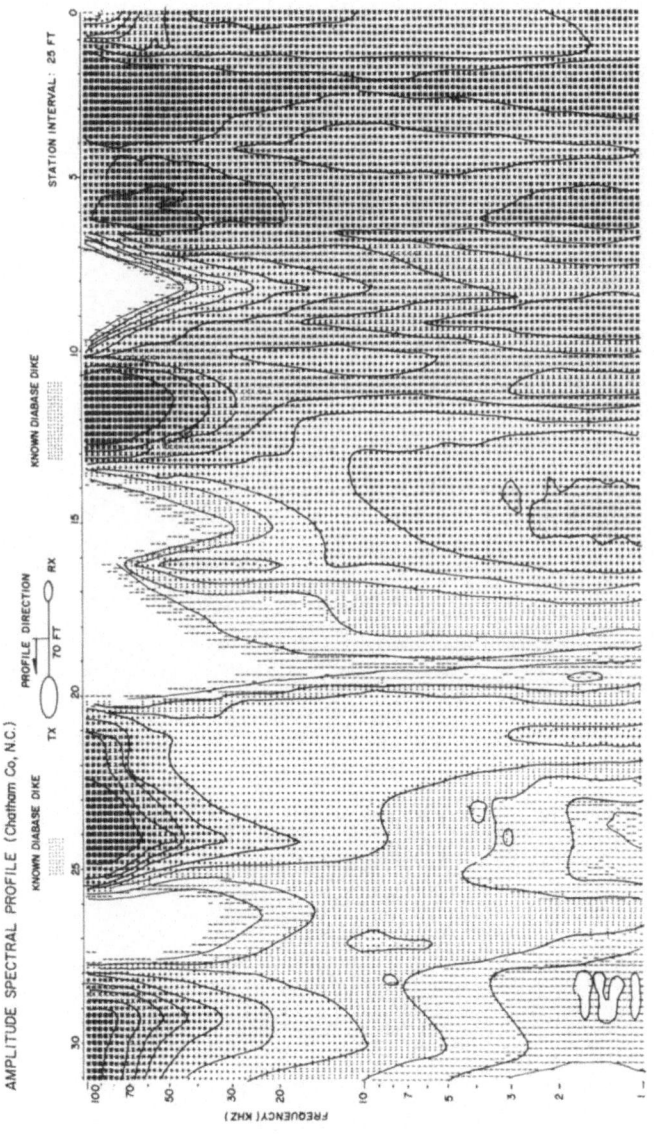

FIG. 12. The same data as in Fig. 11 except that the low-frequency (deep source) variations are removed, thus enhancing the high-frequency (shallow source) anomalies.

I. J. WON

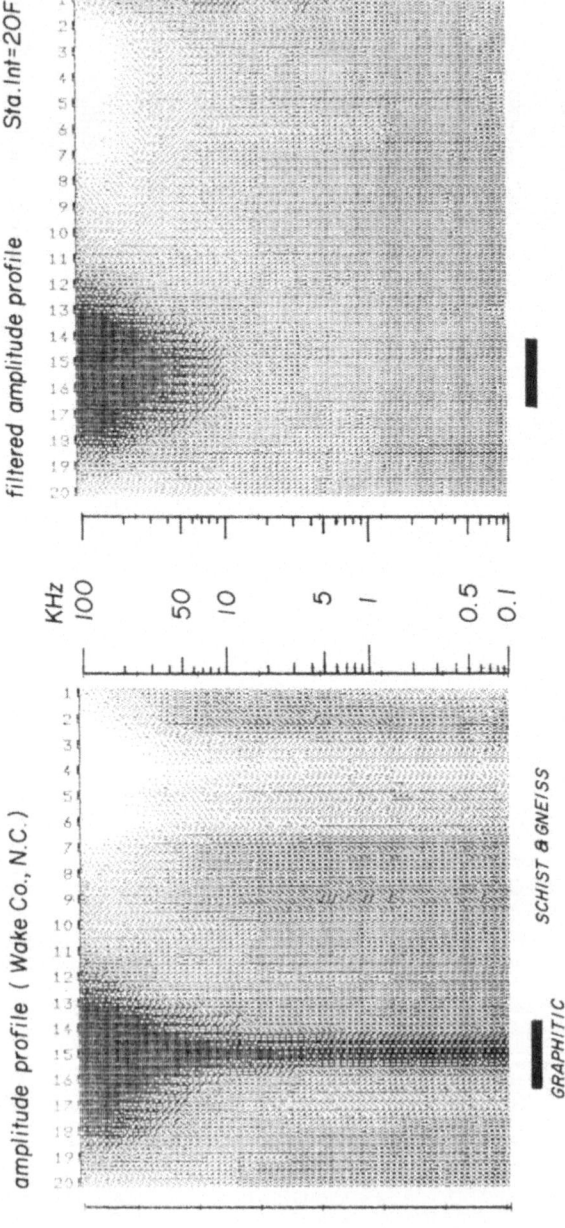

Fig. 13. Amplitude spectral profiles over a known graphite vein.

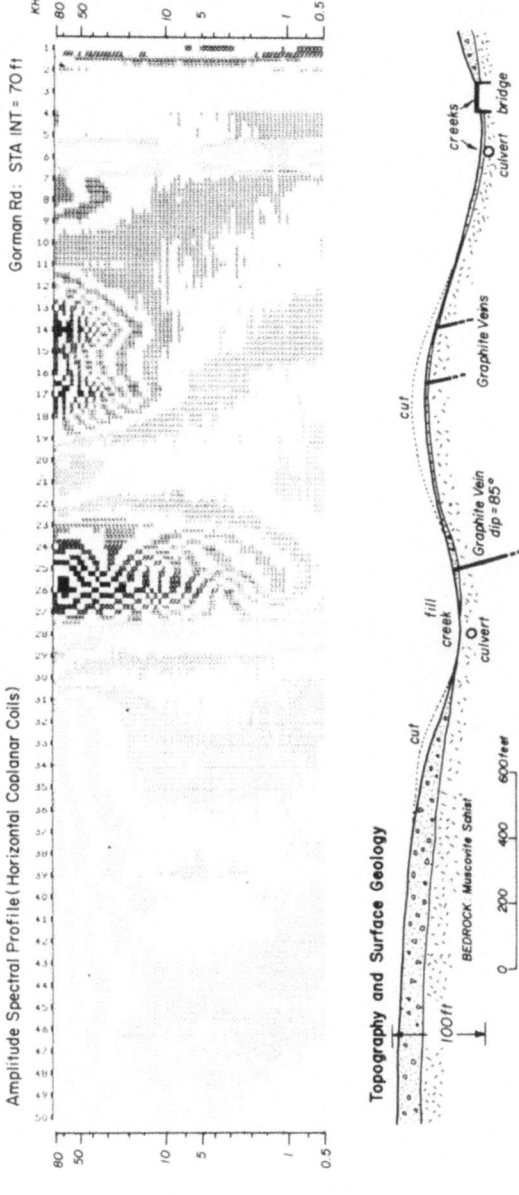

FIG. 14. Amplitude spectral profile from a portion of Gorman Road, south of Raleigh, N.C., along with the available geological data.

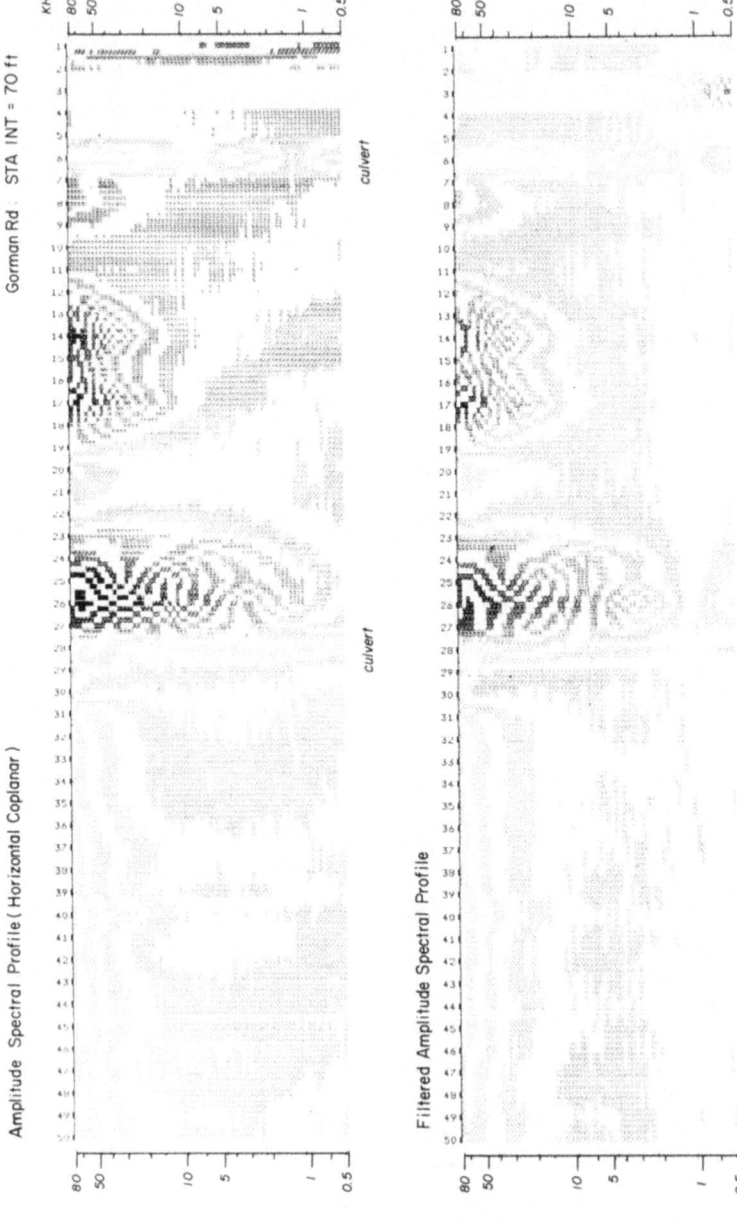

FIG. 15. An enlarged section from a portion of Gorman Road, south of Raleigh, N.C. All known cultural objects are shown in the section. The high-amplitude anomalies near stations 12–18 and 23–27 coincide with graphitic schist formations.

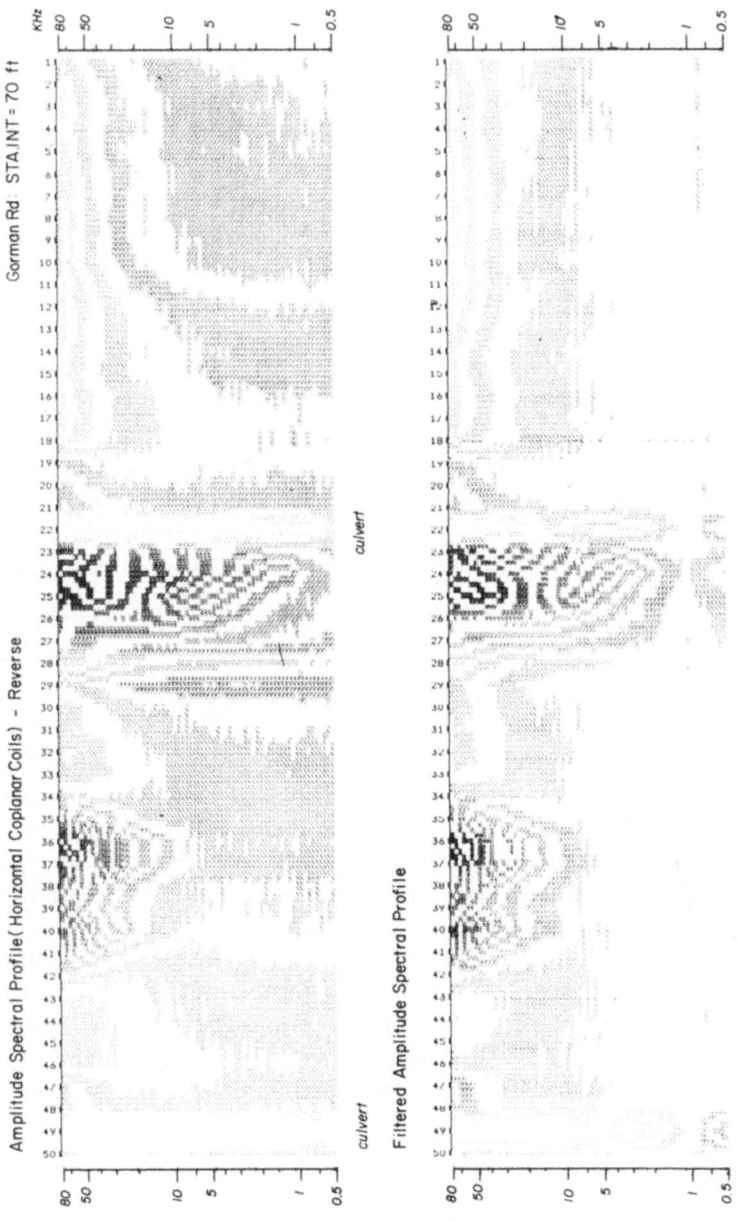

FIG. 16. A reversed profile of the section shown in Fig. 15. This section is made to check reproducibility and effects of vehicle orientation. Since the road is a four-lane highway and the profiles were made along the shoulder, the two profiles shown in Figs 15 and 16 are about 100ft apart.

the road is a four-lane highway and the profiles were made along the shoulder, the two profiles are separated by about 100 ft. The two strong anomalies appearing near stations 14 and 25 on the forward profile and near stations 25 and 36 on the reverse profile are believed to be caused by veins of graphitic schist outcropping in the neighbourhood.

Figure 17 show the same area as in Fig. 15 with the corresponding phase spectral profile. The erratic phase spectra between 900–1100 Hz and 65–80 kHz are due to instrumental errors caused by a sudden change in analog output voltage of the phasemeter whenever the phase of the total field changes from $-180°$ to $+180°$ or vice versa. By combining both the amplitude and the phase information, one may compute the in-phase or quadrature components as well.

Finally, Fig. 18 shows a profile obtained by a minimum-coupled configuration: horizontal transmitter and vertical receiver coils. The data area is the same as in Fig. 15. The primary field is generally about 40 dB lower than that of the maximum-coupled configuration causing the noisy background. Again, the strong anomalies near stations 22 and 33 are believed to be caused by veins of graphitic schist outcropping in the vicinity. It appears that the system is not sensitive under this configuration.

6. CONCLUSIONS

A spectral profile displayed in a frequency–distance domain is somewhat analogous to a reflection seismic profile displayed in a time–distance domain. Although the 3-D seismic diffraction theory is far from complete and somewhat more complicated than the EM diffraction theory, the seismic reflection data are often self-evident in terms of the relative geometry of reflecting horizons. It is premature to say that the conversion of frequency–distance EM data into a conductivity–depth section would be as definite as the conversion of a seismic time section into a seismic depth section; the spectral EM theory must be developed much further to reach that stage. However, the theoretical and experimental data suggest that such a self-evident interpretation does exist in the spectral EM profiles.

It should be noted, however, that the state of the art is far from complete. Needless to say, we need further theoretical development in wide-band frequency response of realistic earth models as well as advances in experimental schemes. With further progress in these aspects

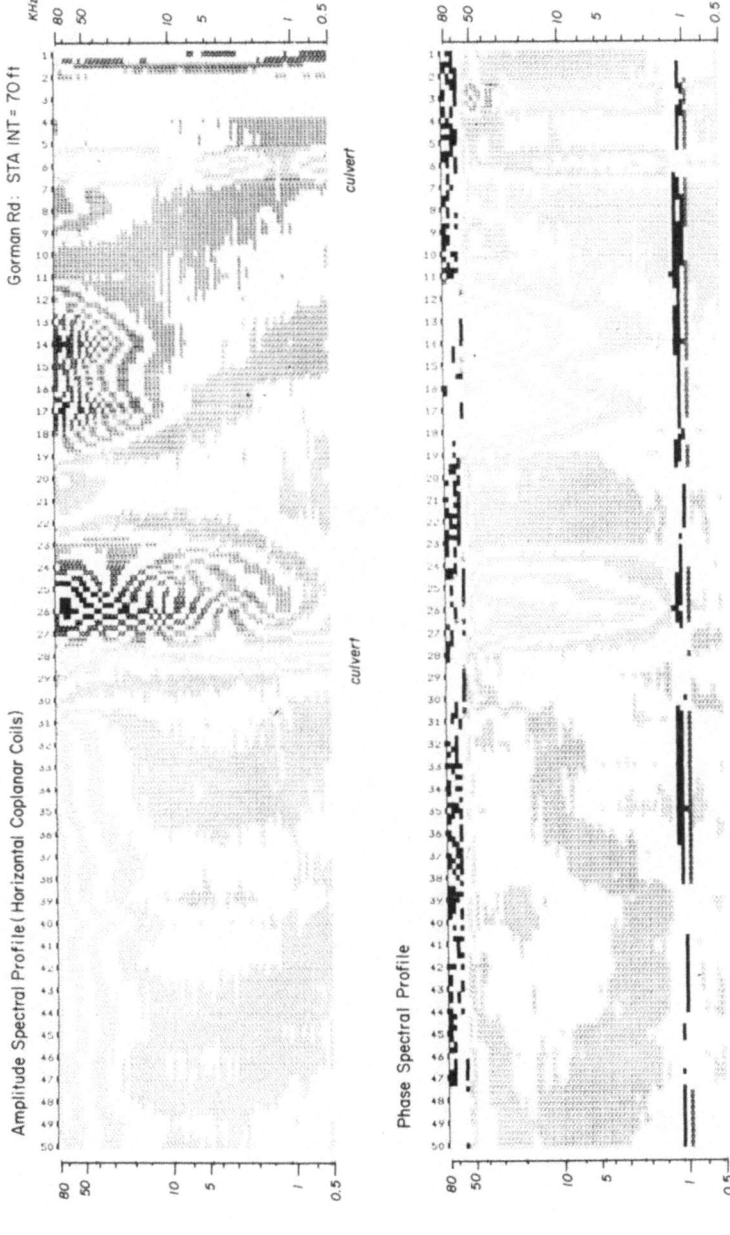

FIG. 17. The same data as in Fig. 15 with phase spectral profile. Contour interval is 0.4°. The erratic phase spectra between 900–1100 Hz and 65–80 kHz are due to instrumental error (see text).

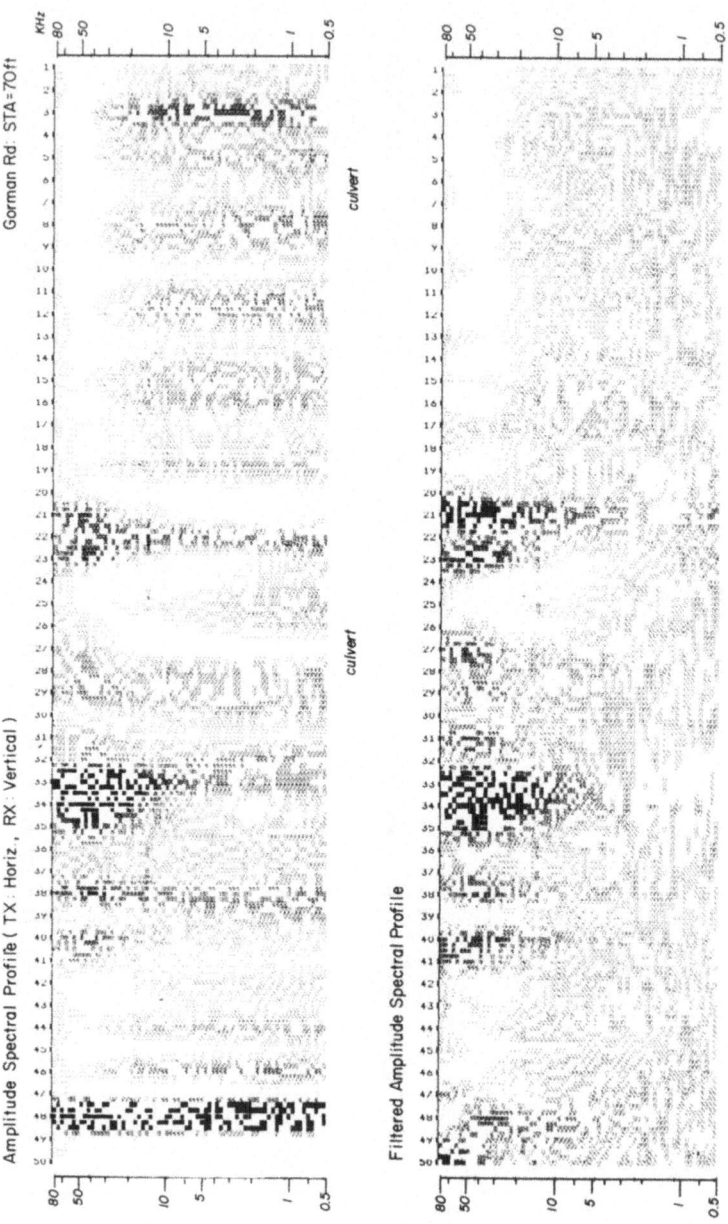

FIG. 18. The same areas as in Fig. 15 measured by a minimum-coupled transmitter and receiver configuration. The primary field amplitude is generally about 40 dB lower than that of the maximum-coupled configuration causing the noisy background.

we may be able to develop the present method into an effective and efficient tool for exploring geological structures and resources.

ACKNOWLEDGEMENTS

The author is grateful to Dr Walter L. Anderson of the US Geological Survey in Denver for making his computer program for computing frequency response of layered earth available to us. This research is supported by the US Army Research Office under contract DAAG29–79–C–0057.

REFERENCES

ANDERSON, W. L. (1974) Electromagnetic fields about a finite electric wire source. US Geol. Serv. Rep. GD-74-041, 205 pp.; NTIS Rep. PB-238-199.

COGGON, J. H. (1971) Electromagnetic and electrical modeling by the finite-element method. *Geophysics* **36**, 132–55.

FRISCHKNECHT, F. C. (1967) Fields about an oscillating magnetic dipole over a two-layered earth and applications to ground and airborne electromagnetic surveys. *Q. Colo. School Mines* **62**(1), 326 pp.

GRANT, F. S. and WEST, G. F. (1965) *Interpretation Theory in Applied Geophysics*, McGraw-Hill, New York, 584 pp.

HOHMANN, G. W. (1971) Electromagnetic scattering by conductors in the earth near a line source of current. *Geophysics* **36**, 101–31.

HOHMANN, G. W. (1975) Three-dimensional induced polarization and electromagnetic modeling. *Geophysics* **40**, 309–24.

KELLER, G. V. and FRISCHKNECHT, F. C. (1966) *Electrical Methods in Geophysical Prospecting*, Pergamon Press, New York, 517 pp.

LAJOIE, J., ALFONSO-ROCHE, J. and WEST, G. F. (1975) Electromagnetic response of an arbitrary source on a layered earth: a new computational approach. *Geophysics* **40**, 773–89.

PARRY, J. R. and WARD, S. J. (1971) Electromagnetic scattering from cylinders of arbitrary cross-section in a conductive half-space. *Geophysics* **36**, 57–100.

RIJO, L., WARD, S. H., HOHMANN, G. W. and SILL, W. K. (1977) On interpretation of broad-band electromagnetic data. Presented at the 47th Annual International Meeting of the Society of Exploration Geophysicists, Calgary, 20 Oct. 1977.

RYU, J., MORRISON, H. F. and WARD, S. H. (1972) Electromagnetic depth sounding experiment across Santa Clara Valley. *Geophysics* **37**, 351–74.

WAIT, J. R. (1962) *Electromagnetic Waves in Stratified Media*, Pergamon Press, New York, 372 pp.

WARD, S. H. (1967) Electromagnetic theory for geophysical applications. In *Mining Geophysics*, Vol. 2, Society of Exploration Geophysicists, Tulsa.

WARD, S. H., PRIDMORE, D. F., RIJO, L. and GLENN, W. E. (1974) Multispectral electromagnetic exploration for sulphides. *Geophysics* **39**, 662–82.

WARD, S. H., PRIDMORE, D. F. and RIJO, L. (1977) *NSF Workshop in Mining Geophysics*, University of Utah, 309 pp.

WON, I. J. (1980) A wide-band electromagnetic exploration method: some theoretical and experimental results. *Geophysics* **45**, 928–40.

WON, I. J. and KUO, J. T. (1975) Representation theorems for multiregional electromagnetic diffraction problem, Part I: Theory. *Geophysics* **40**, 96–108. Part II: Applications. Ibid. **40**, 109–19.

Chapter 3

THE MAGNETIC INDUCED
POLARISATION METHOD

H. O. Seigel

Scintrex Ltd, Concord, Ontario, Canada

and

A. W. Howland-Rose

Scintrex Pty Ltd, Killarney Heights, New South Wales, Australia

SUMMARY

The magnetic induced polarisation or MIP^R method determines the variation of the induced polarisation and resistivity of the earth through measurements of the magnetic field associated with galvanic current flow in the earth, rather than the electric field, as in the traditional IP or EIP method. There are important differences in field practice, mathematical theory and field results between the MIP and EIP methods. For example, the MIP method is insensitive to horizontal layering in the earth and reflects only lateral variations in its electrical properties. It also provides the ability to detect the presence of bodies of anomalous electrical properties even through a highly conducting surface layer. For this reason the MIP method has found its primary application in regions of highly conducting (e.g. saline) overburden or weathered rock, such as in Australia. MIP responses tend to be more complex and varied in pattern than are normally encountered in EIP measurements. For example, polarity reversals are the rule in the former but are rarely encountered in the latter. MIP employs high-sensitivity component magnetometers as basic sensors. These are small in size relative to the length of the electric dipole sensors normally employed in EIP and therefore provide relatively higher geometric resolving power.

MIP^R is a registered trademark of Scintrex Ltd.

1. INTRODUCTION

The induced polarisation (IP) method has been actively and successfully applied in mineral explorations since its first use in 1948 by the Newmont group (Wait, 1959). It is accepted and employed as a basic electrical prospecting method, particularly for the detection of sulphide-related ore deposits of low intrinsic conductivity, such as porphyry coppers and bedded lead–zinc deposits. For a recent summary of the general subject of IP the reader may be referred to Sumner (1979).

The traditional method of measuring IP effects, whether employing the so-called 'frequency domain' or 'time domain' approach, has as a common feature the measurement of one or more characteristics of the electric fields associated with the galvanic passage of current in the earth. In brief (Fig. 1), current of a predetermined time-varying waveform is caused to pass into the ground through a pair of ground points (current electrodes C_1 and C_2) and the resultant earth voltages are measured across another pair of ground points (potential electrodes P_1 and P_2), suitably placed relative to the current electrodes. The IP characteristics of the earth are determined by detecting differences in time (phase) or shape (or both) between the waveforms of the applied current and the earth voltages. For the purposes of what is to follow, it is convenient for us to refer to this traditional IP method as the 'EIP' method since it relies on the electric fields for measurement.

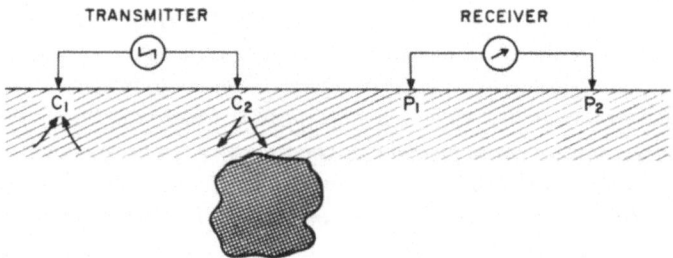

Fig. 1. Traditional IP (EIP) measurement scheme.

The EIP method has served the mining industry well in the past three decades and has many mineral discoveries to its credit (see, e.g., Seigel, 1971). Nevertheless, in common with all mineral exploration methods, it has its limitations. Perhaps the most serious EIP limitation is associated with the 'masking effect' imposed by extreme resistivity contrasts in the earth. For example, should there be a relatively highly (ionic) conducting layer lying between a geological target and the ground surface, the electric field indications from such a target may readily be short-circuited, thus masking its presence. Unfortunately, such a geological environment often exists in areas where there is an overburden or weathered layer whose conductivity is one or more orders of magnitude higher than that of the underlying rock. In such areas this limitation, if recognised at all, has been accepted with resignation by explorationists.

However, starting on an experimental basis in 1971 and progressing through to full-scale field use, measurements have been made of the *magnetic* fields associated with galvanic current flow, with the objective of revealing areas of anomalous IP characteristics. This approach, designated the magnetic induced polarisation or MIP[R] method (Fig. 2), has characteristics, both theoretical and practical, which differ in important aspects from EIP. Perhaps the most important difference lies in the ability of the magnetic fields to penetrate conducting surface layers without serious attenuation. A second important characteristic of MIP is that it is insensitive to the electrical properties (IP or resistivity) of a horizontally stratified earth made up of layers each of which has uniform physical properties throughout. As a consequence, the MIP method is responsive only to lateral variations in the electrical properties of the earth, and is similar to the gravity method in this respect. The EIP method, on the contrary, provides quantitative values of these electrical properties of the earth, albeit distorted by geometric effects and averaged, regardless of the manner in which they are distributed throughout the earth.

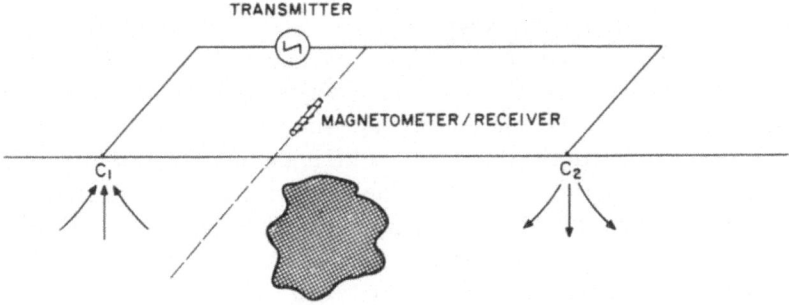

FIG. 2. Typical MIP measurement scheme.

MIP measurements reflect the magnetic fields due to polarisation and ohmic currents, both within and outside of buried polarisable bodies and, as such, tend to produce more complex and varied patterns than are normally encountered in EIP measurements. For example, polarity reversals are the rule in MIP, but are rarely encountered in EIP.

Individual MIP measurements are essentially 'point' measurements, since the physical dimensions of the magnetic sensors used are less than 0.6 m. EIP measurements, on the contrary, usually sum the earth voltages over dipole lengths which are in excess of 50 m. Thus, the potential resolving power of MIP measurements is greater than that of EIP measurements.

These and other idiosyncrasies of the MIP method will be described more fully in the following sections. On the whole it may be concluded that EIP and MIP are so different in nature as to be considered complementary rather than competitive approaches to IP measurements.

One of the basic quantities measured in the practice of MIP is the magnetic field due to the ohmic (steady-state) current flow in the earth. This is the source measurement of the 'magnetometric resistivity method' (MMR) which has been employed as a distinct method by Edwards (1974).

Much of the material presented herein was first published in articles by Seigel (1974) and Howland-Rose *et al.* (1980, Parts I and II).

2. THEORETICAL DEVELOPMENT

In the discussion which follows, the MKS system of units will be employed. In addition, for simplicity, we shall ignore electromagnetic induction effects related to the finite conductivity of the earth. This assumes that the effective induction parameters are kept sufficiently low to be neglected—which is, of course, not true in all cases.

The basic mathematical theory of the MIP method was first presented by Seigel (1974). The magnetic field due to current flow in the earth will consist of two parts: firstly the 'primary magnetic field', viz. that which is due to the ohmic current flow, possibly distorted by conductivity variations in the earth; and secondly the 'secondary magnetic field' which is caused by the polarisation characteristics of the ground.

The primary vector magnetic field, $\mathbf{H}_p$, may be expressed by the Biot-

Savart law:

$$\mathbf{H}_{\mathrm{p}} = \frac{1}{4\pi} \int \int_V \int \frac{\mathbf{j}_{\mathrm{p}} \times \mathbf{r}}{r^3} \, dv \qquad (1)$$

where V is the entire volume through which the primary current (local current density j_{p}) flows and r is the distance from the observation point to the elemental volume dv.

If we have axial symmetry of physical properties, as would prevail when current is passed from a current electrode into a horizontally stratified earth, we may usefully employ Ampère's law to determine that on the surface of the earth the magnetic field due to current flow has only a horizontal component, H_{p}, given by

$$H_{\mathrm{p}} = \frac{I}{4\pi r} \qquad (2)$$

in amperes per metre, r being the distance from the electrode in metres and I the current in amperes. H_{p} is perpendicular to the line joining the observation point to the current electrode.

The magnetic field, due to current in the cable feeding the current electrode, has only a vertical component, provided that the cable is laid out on the flat surface (earth's surface) on which the measurements are being made. Thus, by measuring the horizontal magnetic field which is largely orthogonal to the line joining the current electrode to the point of measurement, one observes primarily the magnetic field due to subsurface current flow.

It should be noted that eqn (2) is valid regardless of the stratification and thus is insensitive to the nature of the stratification, provided only that it is axially symmetric about the current electrode. The most common geological occurrence fitting this requirement is flat-lying formations, each having its intrinsic, uniform electrical properties.

The presence of polarisation effects in the earth may change the current distribution and therefore the resultant magnetic fields that will be measured. We may express the polarisation magnetic field (or 'secondary' field) in vector form as a perturbation on the primary field (eqn (1)) as follows:

$$\mathbf{H}_{\mathrm{s}} = \Delta \mathbf{H}_{\mathrm{p}} = \frac{1}{4\pi} \int \int \int_V (\Delta \mathbf{j}_{\mathrm{p}} \times \mathbf{r}/r^3) \, dv \qquad (3)$$

However,

$$\Delta \mathbf{j}_{\mathrm{p}} \cong \Sigma_i (\partial \mathbf{j}_{\mathrm{p}}/\partial \sigma_i) \Delta \sigma_i \qquad (4)$$

where σ_i is the conductivity of the ith region in the earth. In the time domain, for example, as was proposed by Seigel (1959),

$$\Delta\sigma_i \cong -M_i\sigma_i \qquad (5)$$

where M_i is the chargeability of the ith region. Thus, we find that

$$\mathbf{H}_s = -\frac{1}{4\pi}\sum_i M_i \iiint_V \frac{\partial \mathbf{j}_p}{\partial \log\sigma_i} \times \mathbf{r}/r^3 \, dv \qquad (6)$$

Similarly, in the frequency domain, if we are measuring a change in magnetic field amplitude with a change in frequency, we find

$$\frac{\Delta\sigma_i}{\sigma_i} \cong -\frac{(\mathrm{PFE})_i}{100} \qquad (7)$$

where σ_i and $(\mathrm{PFE})_i$ are the conductivity and the customary 'percentage frequency effect' of the ith domain in the medium. We then find that

$$\mathbf{H}_s = -\frac{1}{4\pi}\sum_i \left(\frac{\mathrm{PFE}}{100}\right)_i \frac{\partial}{\partial \log\sigma_i} \iint_V \int \frac{\mathbf{j}\times\mathbf{r}}{r^3}\,dv \qquad (8)$$

If the pertinent measurement in the frequency domain is of the out-of-phase component of the magnetic field, relative to the phase of the primary current, then we may express the conductivity of the ith medium as $\sigma_i\,(1+j\tan\theta_i)$, where θ_i is the phase angle of the conductivity (assumed to be complex due to induced polarisation), and thus

$$\frac{\Delta\sigma_i}{\sigma_i} = \tan\theta_i \qquad (9)$$

so that the quadrature component becomes

$$\mathbf{H}_s = \frac{1}{4\pi}\sum_i \tan\theta_i \frac{\partial}{\partial \log\sigma_i} \iint_V \int \frac{\mathbf{j}_p\times\mathbf{r}}{r^3}\,dv \qquad (10)$$

Thus, regardless of the quantity being measured, the calculation of the MIP response of an earth consisting of a number of discrete domains, each with its own conductivity and induced polarisation property (chargeability, PFE or phase angle), is made as follows:

(a) solve the corresponding steady-state current flow case (Laplace's equation) to determine the distribution of $\mathbf{j}_p$, the primary current flow through the earth;

(b) integrate $\mathbf{j}_p \times \mathbf{r}/r^3$ over the whole earth and then differentiate the integrals with respect to $\log\sigma_i$; then

(c) multiply each derivative by its corresponding IP characteristic, and sum for all domains.

Of course, the order of integration and differentiation may be reversed. For all but very simple geometries it is impossible to express these quantities in closed form, and computer assistance is required. Analogue models may also be employed, but these are not usually convenient from a physical property standpoint, as there is no equivalence of conductivity and frequency, for example, as there is in EM modelling.

When making EIP measurements it is customary to measure the ratio of secondary (polarisation) field to the corresponding primary (steady-state) field to obtain the IP parameter (chargeability, PFE or phase shift). Similarly, in MIP, if we normalise any of eqns (6), (8) or (9) with respect to H_p, we obtain, for example (eqn (6), employing scalar notation for a particular component):

$$\mathbf{H}_s/\mathbf{H}_p = M_a = -\sum_i M_i \frac{\partial \log H_p}{\partial \log \sigma_i} \tag{11}$$

The quantity $\partial \log H_p/\partial \log \sigma_i$ now appears as the MIP weighting factor for the ith domain.

In EIP (Seigel, 1959) the corresponding weighting factor is given by $\partial \log \rho_a/\partial \log \rho_i$, where ρ_a is the apparent resistivity of the measurements and ρ_i is the intrinsic resistivity of the ith domain. It was shown for EIP (Seigel, 1959) that

$$\sum_i \frac{\partial \log \rho_a}{\partial \log \rho_i} = 1 \tag{12}$$

i.e. the sum of the EIP weighting factors is unity. Similarly, it may be shown by means of a Taylor expansion (Howland-Rose et al., 1980, Part I) that in MIP

$$\sum_i \frac{\partial \log H_p}{\partial \log \sigma_i} = 0 \tag{13}$$

i.e. that the sum of the MIP weighting factors is zero.

We may derive from eqn (13) the conclusion that a medium of constant chargeability cannot give rise to an MIP response, regardless of the variations of its conductivity from place to place within it. Also, eqn (13) leads us to expect as many positive as negative apparent chargeabilities in different regions of space.

As a simple illustration of this bipolar nature of MIP responses, one may refer to the example shown in Fig. 3. Here we see, in section form, the current flow pattern that might be associated with a polarisable body immersed in a medium which is non-polarisable and of somewhat lower conductivity than the body. The body lies within the current flow introduced by two current electrodes at the surface. The IP effect of this body on the galvanic current flow in its vicinity may be represented by the superposition of a second (sourceless) current distribution, which we may call the 'polarisation current'. The steady-state current flow, ignoring EM and IP effects, but including any disturbance due to the electrical conductivity contrasts in the medium, may be designated as the 'ohmic' current flow.

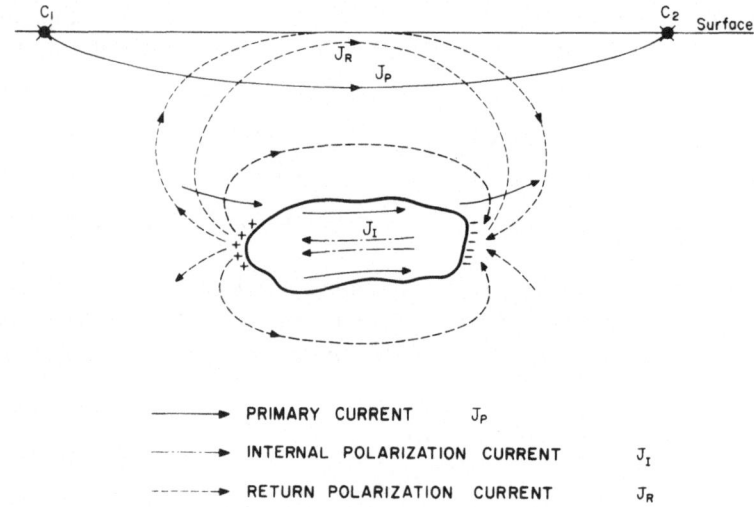

FIG. 3. Ohmic and polarisation current around a buried polarisable body with the current electrodes C_1 and C_2 at the surface (vertical section).

Thinking in terms of time-domain measurements, the sense of the polarisation current is in opposition to the ohmic current flow within the body, as well as 'upstream' and 'downstream' from it. In other regions of space the sense of the polarisation current is the same as that of the ohmic current. The polarisation current within the body may be called the 'internal' polarisation current, and that external to the body the 'return' polarisation current. The return current flow is effectively that which would be due to a bipolar charge distribution, with positive charges at the points of ohmic current entry into the body and negative charges at the points of ohmic current exit from the body.

Figure 4 is a plan view of the horizontal magnetic field component H_s, at right angles to the line joining the current electrodes, due to the polarisation current flow generated by the polarisable body shown in Fig. 3. The sense of H_s is taken relative to that of the same magnetic field component due to the ohmic or steady-state current flow. H_s is thus seen to be negative in the region over the body which is dominated by the internal polarisation current, and positive on its flanks, orthogonal to the main ohmic current flow direction.

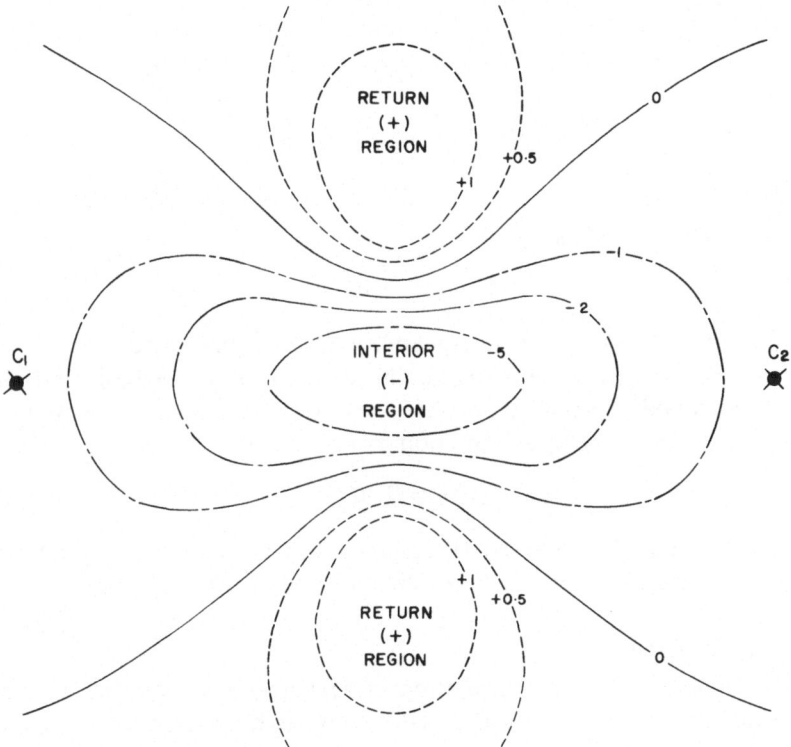

FIG. 4. Horizontal polarisation magnetic field component H_s orthogonal to the line joining the current electrodes (plan view).

These reversals of sign are common in MIP but are rarely encountered in EIP measurements. In EIP one usually measures the electric fields exterior to the target body (normally at the surface), i.e. associated with return currents. These, as we have noted, are generally in the same sense as the primary electric field in most regions of space.

A further consequence of the differences in the weighting factors for EIP and MIP is that the observed anomalous MIP spectral response parameters will be more closely related to the intrinsic spectral response parameters of the IP source. For example, if we are dealing with a two-component earth, where component 1 is the uniform earth and component 2 is a target body of anomalous IP, we may express the apparent chargeability M_a as

$$M_a = B_1 M_1 + B_2 M_2 \qquad (14)$$

where $B_1 M_1$ and $B_2 M_2$ are the weighting factors of the first and second components, respectively. Now in the case of EIP, since $B_1 + B_2 = 1$ (eqn (12)),

$$M_a = M_1 + B_2 (M_2 - M_1) \qquad (15)$$

and if $M_2 \gg M_1$ for all measuring times—which is normally true—then we may rewrite this as

$$M_a = M_1 + B_2 M_2 \qquad (16)$$

that is, the observed EIP chargeability spectral parameters (or decay curve form) involve a mixture of those of the host medium and the anomalous body. In the case of MIP, however, since $B_1 + B_2 = 0$, we may find that, to the same approximation,

$$M_a = B_2 M_2 \qquad (17)$$

that is, in MIP the observed IP response will have spectral parameters which are closer to those of the anomalous body itself.

2.1. Dike Model

As an illustration of the calculation of MIP response patterns we may consider the case of a vertical dike-like tabular body. It is assumed that the current electrodes are placed on the mid-line of the dike with a separation of $2L$ The dike is assumed to have a resistivity ρ_2 and PFE of 10% over the frequency span employed, and to be embedded in rocks of resistivity $\rho_1 = 10\rho_2$ and zero PFE. The dike width is taken to be $0.4L$. This body may simulate a zone of disseminated sulphides in a shear zone, for example.

Figure 5 shows the primary magnetic field over the centreline of the current electrodes, at right angles to the dike. This field is expressed in three different fashions which may be employed from time to time. These

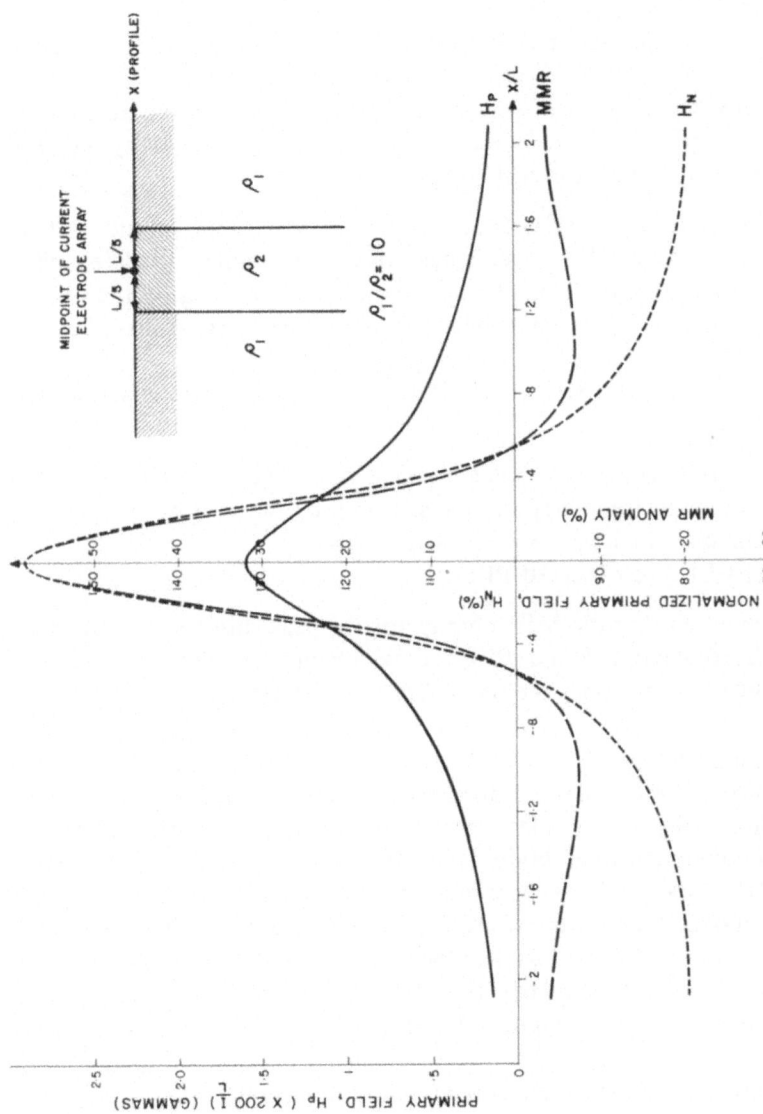

FIG. 5. Primary magnetic field over a conductive dike ($\rho_1/\rho_2 = 10$).

are:

(a) H_p, normalised by division by the anticipated uniform earth field at the centre of the line;

(b) H_N, normalised by division at each point by the calculated uniform earth field at that point; and

(c) MMR, H_p normalised further by *subtraction* of the anticipated uniform earth field at each point. This latter terminology is due to Edwards and Howell (1976).

All the representations show more or less the same general trends, viz. an abnormal increase of magnetic field over the conducting dike and a decrease on its flanks. This of course reflects the robbing by the dike of current which would normally have flowed in the wall rocks on either side of the dike.

Figure 6 shows the MIP effect of the dike, plotted in two characteristic ways, viz.:

(a) H'_s, the secondary or polarisation field, i.e. the change in field with the change in frequency normalised only by division by the primary current which passes in the array; and

(b) $(PFE)_a$, the observed PFE.

It will be noted that the MIP characteristics peak directly over the dike, giving rise to negative H'_s and $(PFE)_a$, typical of interior polarisation current, and positive or return current flow effects on the flanks.

2.2. Confined Bodies

Geological targets of limited dimensions are, of course, of great interest for exploration. The simplest model for such targets is the sphere, or roughly equidimensional body. It has been shown (Seigel, 1974) that the peak MIP response of a sphere in a uniform energising field, at the point directly above its centre, drops off as the inverse square of its depth below the surface. This is one power less than the attenuation of the EIP (electric field) response with depth. Similarly, the MIP response of horizontal, pipe-like bodies, with long strike length, will follow an inverse first power law of attenuation with depth.

For tabular, disc-like bodies, not an uncommon model for some sulphide-rich, strata-bound deposits, there is a particular advantage in passing current longitudinally, i.e. along their long dimension. When this is done, in the case of a relatively highly conducting body, it becomes feasible to concentrate considerable amounts of current within the body, and thus

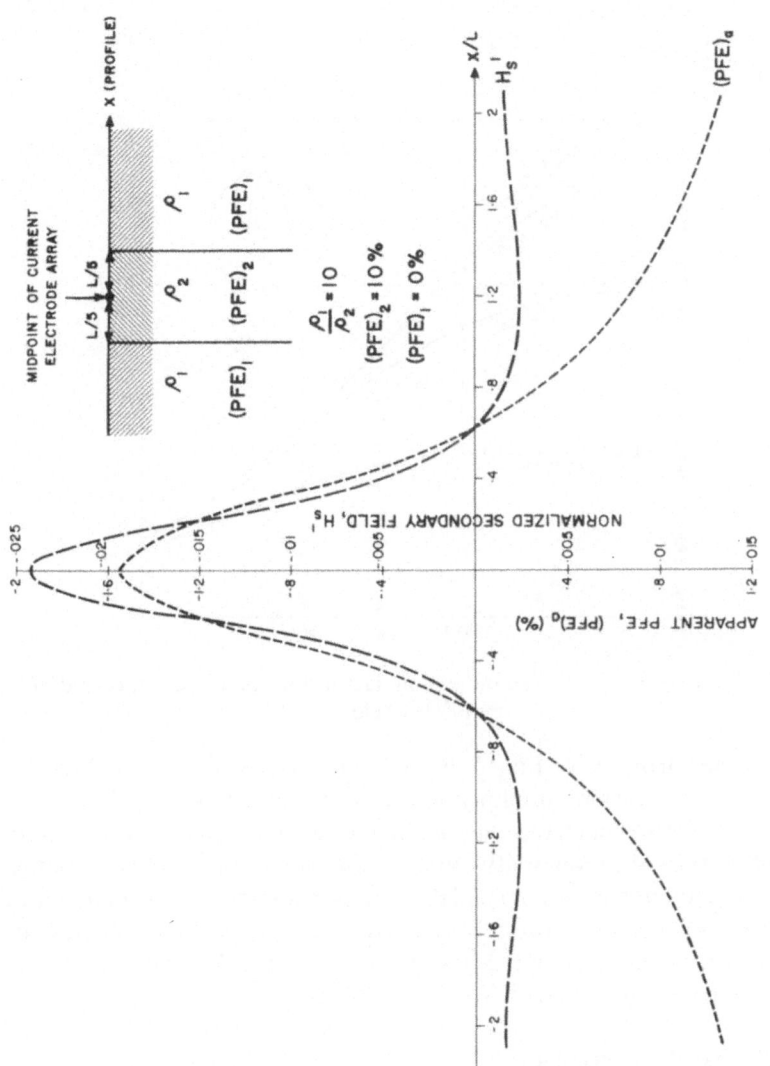

FIG. 6. MIP effect over a polarisable conductive dike ($(PFE)_2 = 10\%$, $\rho_1/\rho_2 = 10$).

greatly to enhance the body's response over what it would be if current were to be passed *across* its long dimension. Figure 7 shows the interior current amplification factor for a disc-like model, based on an oblate spheroid (long axis a, short axis b, interior conductivity σ_1, in a medium of conductivity σ_0).

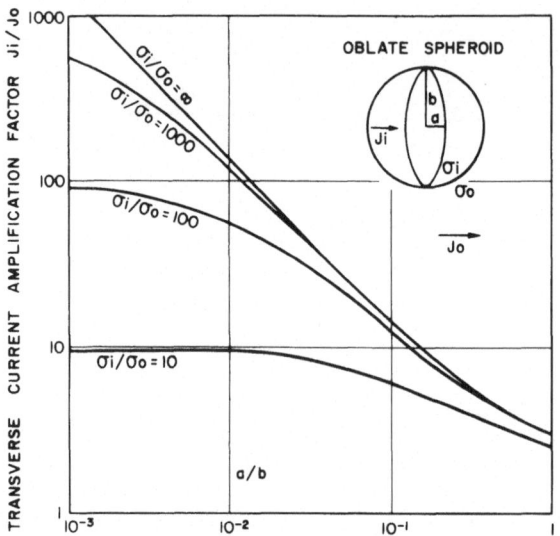

FIG. 7. Transverse current amplification factor for an oblate spheroid in a uniform field.

It is noteworthy from Fig. 7 that for large values of the conductivity ratio σ_1/σ_0, the current amplification factor is limited only by the ratio of the axes, b/a. Concentration factors of 100 or more may readily be anticipated for typical conductivity ratios in disc-like, sulphide-rich orebodies. In addition, there may still be an IP effect from a rather highly conducting sulphide body (i.e. an appreciable change of current with change in body conductivity for a high conductivity body), provided that the ratio of axes exceeds the ratio of conductivities.

2.3. The Masking Problem

The masking effect of overlying conducting horizons on the electric field response of underlying targets is a very real problem in EIP measurements. Most commonly the problem stems from a surface layer of overburden which is much more highly conducting than the underlying rocks. Less well appreciated is the effect of a conducting layer which is itself buried but lies above the target horizon. To determine the relative effects of an overlying conducting layer on the electric and magnetic field at the surface due to a disturbing body, we have resorted to a scale model. The scale model, selected to approximate an elongated, flat-lying anomalous body, is a line current element terminated by two equal opposite point current sources, one at each end. This is a simplified representation of the current flow pattern of Fig. 3, the line current simulating the interior current flow, while the source and sink represent the return current flow. Together they constitute a closed, sourceless current system.

The quantities measured in this test were: (a) the horizontal magnetic field component at the surface perpendicular to the current line, simulating MIP measurements; and (b) the electric field component at the surface parallel to the current line, simulating EIP measurements, both being measured along the mid-section of the body. Measurements of both quantities were made for three cases, viz.

1. The body lying in a medium of uniform resistivity.
2. A flat-lying conducting layer, extending from the surface halfway down to the level of the anomalous body.
3. A similar conducting layer, but buried under an equal thickness of material of resistivity equal to that of the underlying medium.

The resistivity of the conducting layer was made to be 65 times less than that of the remainder of the medium, which is within the range often encountered. Figure 8 shows the electric field for all three cases. The electric field is seen to be reduced 30-fold in case 2 and 25-fold in case 3 from that of case 1. In effect, this would reduce the surface EIP response of the anomalous body to negligible levels. Figure 9, however, shows that the magnetic field is essentially unaffected in either case 2 or 3 by the overlying conducting layer. This might have been expected from our earlier remark that a horizontally stratified earth has no effect in the surface magnetic field associated with current flow from a point source.

These results illustrate one of the most important attributes of MIP, viz. its ability to detect electrically anomalous bodies even through a highly conducting overlying layer. This is particularly important in areas of highly saline overburden, e.g. in Western Australia. Of course, this illustration shows only how the signals from the body are affected by conductivity contrasts in the host medium. A further problem lies in the attentuation, by the same conducting layers, of the energising current at the body location. This is normally established through a pair of current

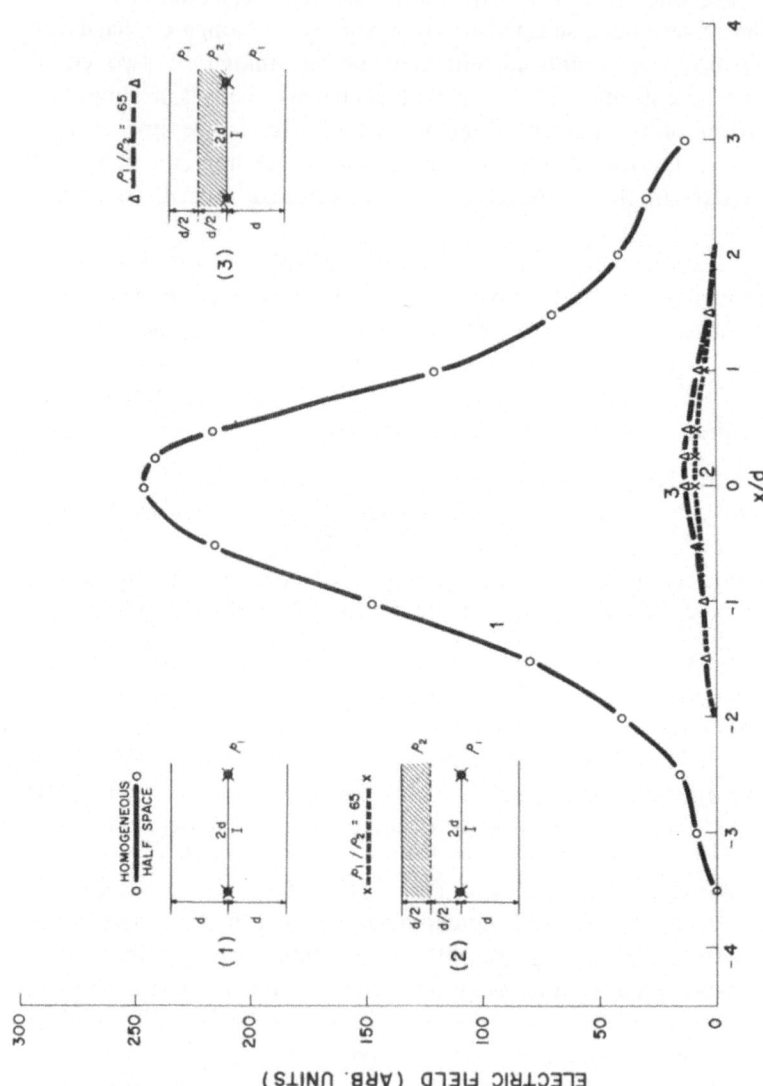

FIG. 8. Effect of conductive layer on electric field (EIP) results. 1, Homogeneous half-space; 2, conductive layer at the surface; 3, conductive layer buried.

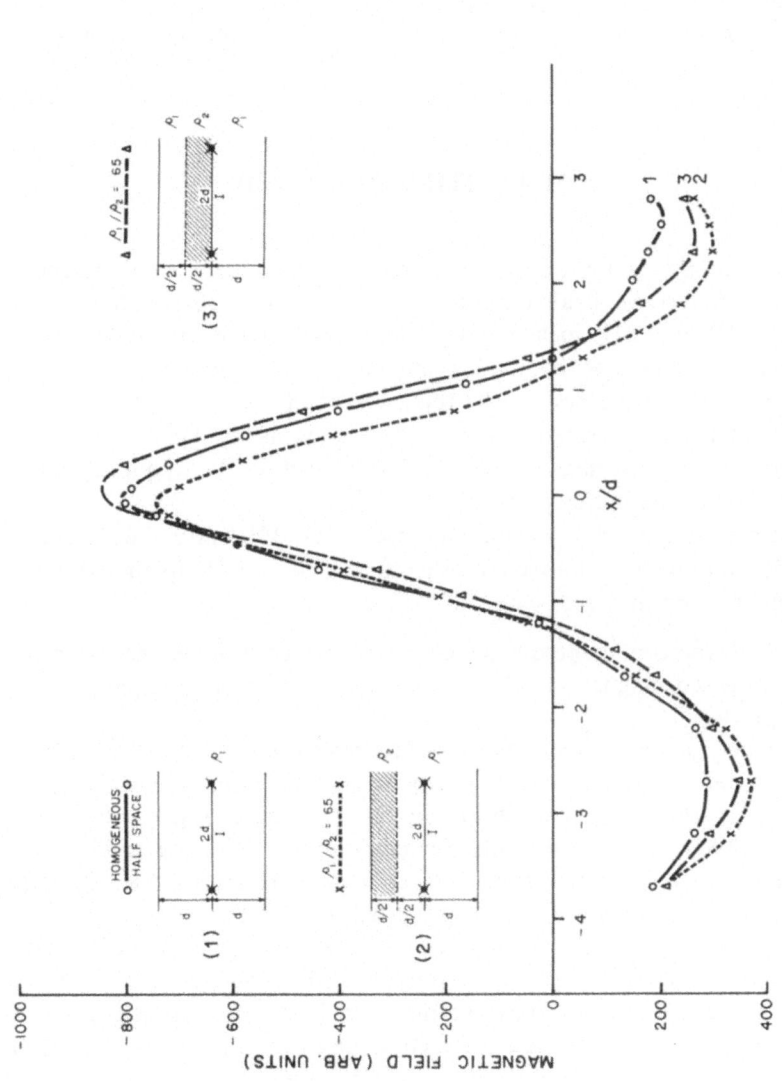

FIG. 9. Effect of conductive layer on magnetic field (MIP) results. 1, Homogeneous half-space; 2, conductive layer at the surface; 3, conductive layer buried.

electrodes at the surface. However, the attenuation of the energising current by the conducting layer will be the same for EIP as for MIP, so that our present model has properly demonstrated the *relative* EIP and MIP responses at the surface of a buried body of anomalous electrical properties.

3. QUANTITIES MEASURED

Aside from the fact that the basic MIP measurement is of magnetic (rather than electric) field components associated with current flow in the earth, MIP and EIP measurements may be expressed in terms of the same parameters and be made with the same equipment. The only difference in measurement is the replacement of the grounded electrode dipole input by a sensitive component magnetometer. To be useful, the magnetometer must have an intrinsic noise level (in the frequency range of interest) in the milligamma region.

The precise parameters being measured in an MIP survey will depend on the equipment employed in the survey. If it is frequency-domain equipment, one may measure:

1. *PFE* (percentage frequency effect), i.e. the percentage change in the observed magnetic field with a change in operating frequency. It is plotted in %.
2. *Phase angle* between the observed magnetic field and the current applied to the ground. It is plotted in degrees or milliradians.
3. *RPS* (relative phase shift): this is the phase angle (time shift) between two harmonically related components of the applied current, as is obtained directly by a Scintrex receiver (IPRF-2) often used for MIP purposes. It is plotted in degrees.
4. H_p, the 'steady state' magnetic field amplitude measured at one frequency. This quantity is rarely plotted *per se*, but is operated on in one of two ways so as to remove the normal variation of H_p with location, assuming that the earth is uniform in physical properties. The resultant quantities, often plotted, are:
   (a) H_N, the normalised primary magnetic field obtained by dividing H_p by H_p', the latter being the value of H_p calculated for the uniform earth for the same location. It is often expressed in %.
   (b) *MMR* (magnetometric resistivity; Edwards and Howell, 1976) $= (H_p - H_{p0})/(H_{p0})$, where H_{p0} is the (uniform earth) pre-

dicted primary field at the point midway between the current electrodes. It is also expressed in %.

When working in the time domain the IP quantity measured is usually the 'chargeability', obtained from a series of mean values of the transient decay curve, averaged over predetermined time gates. These are normalised with respect to H_p and expressed as:

5. M_i (the chargeability corresponding to the ith gate), in millivolts per volt, or 'mils'.
6. The value of H_p is measured during the current on time and is normalised, as in the frequency domain, to produce H_N or MMR.

Whereas these MIP parameters may appear wholly natural to an EIP worker, in fact there are two basic differences. The first is that none of these parameters are actual electrical properties of the medium, averaged in some way, as has been pointed out above. They reflect only lateral variations of electrical physical properties of the medium. The second difference lies in the practice of normalisation with respect to H_p or H_{p0} of all quantities. H_p and H_{p0} are the observed and calculated primary magnetic fields due to current flow in the earth. If one were to keep the current constant and change the separation between the current electrodes, but leave the measuring point at the same place (scaled up or down) relative to the electrodes, the values of H'_p and H_{p0} would be found to change inversely as the first power of the separation. However, the actual perturbation of the IP and steady-state (ohmic) fields due to a body of anomalous electrical properties is governed by the amplitude of the *electric field* in its vicinity. This electric field, in turn, varies inversely as the square of the electrode separation, other things being equal. Thus, if we normalise all MIP quantities with respect to H_p or H_{p0}, we tend to distort the resultant parameters. For example, anomalies which occur well away from the line joining the current electrodes tend to be exaggerated by division by H_p or H'_p and de-emphasised through division by H_{p0}. For this reason, one often plots MIP data in two ways, one being the basic IP parameter such as PFE, RPS and M, and the other being in the form of the anomalous secondary field, as follows: $H_s =$ the same parameter $\times$ H_p/I, in milligammas per ampere, where I is the appropriate primary current in the ground, in amperes. In addition one may plot both H_N and MMR, for the same reason.

A more logical normalisation, albeit one not yet employed in practice, might be to use $j(xy)$ in place of I in the calculation of H_s, $j(xy)$ being the mean near-surface current density near the point of measurement.

4. FIELD PROCEDURES

The MIP response of a body of anomalous electrical properties may be improved by proper field procedure. One important factor is to cause the energising current direction to lie primarily along the probable strike direction of the target bodies. These bodies are normally conformable with the regional formations. Thus, it is standard MIP practice (albeit contrary to EIP practice) to align the current electrodes along the regional strike direction, as determined from geological mapping or magnetic surveys, etc.

Figure 10 shows a typical MIP array as used for the production surveying of an area. The electrodes C_1 and C_2 are set out on a line parallel to the regional geological strike direction, a distance $2L$ apart. They are interconnected by a U-shaped loop of cable which is approximately $2L$ on a side. A rectangular area, about L wide by $2L$ long, may usually be surveyed from one specific current electrode set-up of this type. More than one sensor–receiver system may be employed simultaneously to accelerate the rate of survey coverage.

A second fact to consider when trying to improve the response from a target is to adopt a proper scale (L value) for the array. Because the value of H_p is inversely proportional to L, and that of H_s to L^2, i.e. to the current density in the mid-region of the array, there will be, beyond a certain value of L, a progressive reduction in relative signal level (considering H_p as the reference level). The maximum value of L is normally determined by the mean strike length of targets expected to occur in the area. As a rule of thumb, L should not greatly exceed this mean strike length for best detectability. On the other hand, in order that the current density at the depths at which the targets may occur (depth of cover) may be comparable in magnitude to those near the surface (i.e. which produce the bulk of the H_p field) in the central region of the current array, L should be no less than this depth of cover.

One measures the horizontal magnetic field component along the survey line direction, i.e. orthogonal to the line joining the current electrodes. The station interval along these lines may vary from as little as 10 m, for the detail of very shallow bodies, to 100 m or more for deeply buried targets. In general, the station interval should not exceed one-half of the mean depth of burial (or cover) anticipated in the area, in order to obtain adequate resolution of the observed responses. The interline separation should not greatly exceed the mean expected strike length of

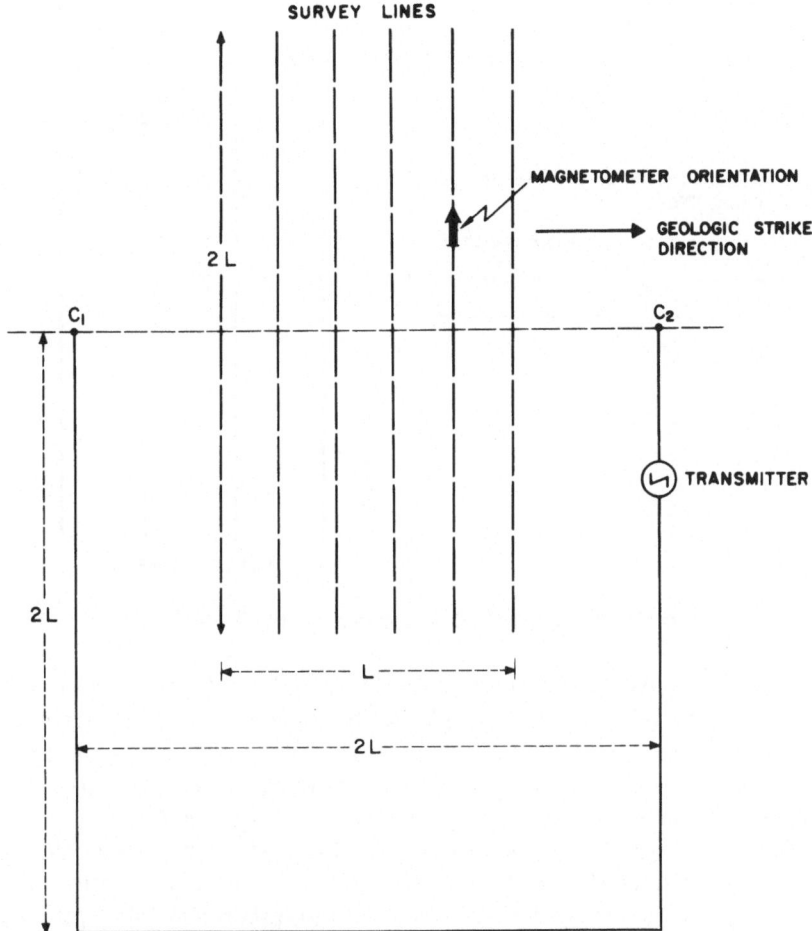

FIG. 10. Typical MIP horseshoe array for reconnaissance surveying.

geological targets of interest in order to have a high probability of detecting them.

A second type of MIP array sometimes employed is shown in Fig. 11. As before, the current electrodes C_1 and C_2 are aligned along the regional strike direction but have a separation $2L$ which is much larger than in Fig. 10. The current cable passes directly, i.e. in a reasonably straight line, between the electrodes. Three or four lines, with spacings as determined above, are surveyed only off one end of the array. Once these

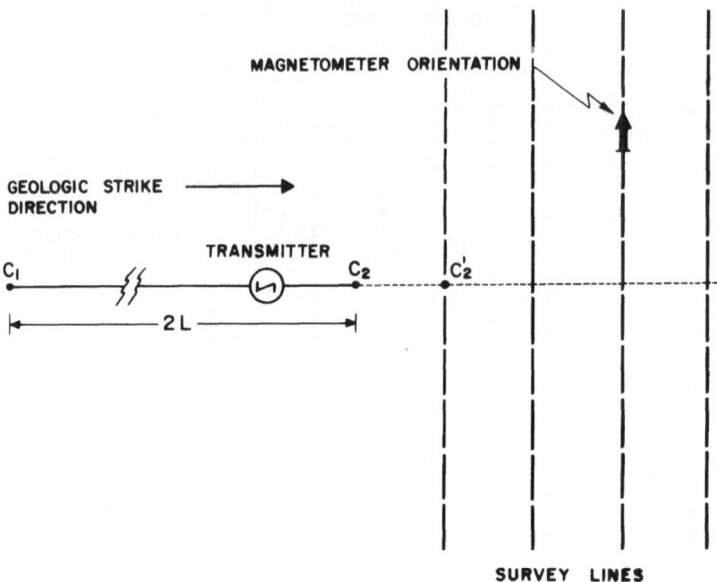

FIG. 11. Typical MIP linear array for detail surveying.

have been surveyed, the current cable is extended and C_2 advances to C_2'. The same pattern of lines, displaced by one line spacing, are then measured. The combined results from such measurements lend themselves well to plotting in pseudo-section form, very much like one standard form of EIP plotting (pole–dipole or dipole–dipole).

The advantage of the standard array of Fig. 10 is high productivity, in part due to enhanced field strength and only one reading per station. The advantage of the array of Fig. 11 is good interpretability for depth and dip of bodies (i.e. better third-dimensional control). It suffers from lower efficiency (more readings per station) and generally lower signal strength, however. For this reason, the array of Fig. 10 is generally used for production or reconnaissance, whereas that of Fig. 11 lends itself well to the detailing of anomalies detected by the production array.

5. FIELD EQUIPMENT

The execution of an MIP survey requires the use of a standard set of EIP equipment, plus the addition of a high-sensitivity vector magnetometer to act as primary sensor, in place of the usual electric dipole (i.e. cable and two electrodes). The specifications required of this magnetometer are as follows:

Noise level: less than 1 milligamma $(m\gamma)/\sqrt{Hz}$
Resolution: better than $1\,m\gamma$
Frequency response: essentially flat, 0–1000 Hz.

The magnetic fields to be measured are usually very small. The value of H_p is commonly less than $0.5\,\gamma$ in the centre of a large array, and we are interested in perturbations of this field which are at least two orders of magnitude lower than this level.

As in EIP measurements, natural magnetotelluric (MT) variations will provide noise background which may, at certain times and places, be a serious impediment to making useful measurements. As a general rule, such MT noise in the frequency range of greatest interest (0–10 Hz) increases progressively with latitude. Higher-frequency noise, associated largely with thunderstorm activity, occurs mainly at middle to low latitudes and, when of local origin, may also have a severe effect on such measurements.

To improve signal-to-noise (S/N) ratios when such problems occur, a number of steps may be taken:

(a) Use of higher power transmitters to provide greater current densities in the ground and thus stronger signals.
(b) Use of narrow-band filters, possible mainly in the frequency domain where phase-lock loops may be employed.
(c) Statistical S/N enhancement resulting from digital stacking and averaging, in both time- and frequency-domain systems. This is a powerful technique, exchanging increased time of measurement for increased power, on an equivalent basis.
(d) Use of a reference magnetometer for cancellation of ambient field noise. This requires selection of a proper location for a fixed reference magnetometer so that it has the same orientation as the moving magnetometer but is not subject to magnetic fields from current flow in the ground due to our current array. The signal from this reference magnetometer is mixed, in series opposition, to the moving magnetometer, using a hard-wired or radio link for transmission.

As in EIP, the choice of MIP measuring in the time domain or frequency domain is a matter of S/N. When the former may be used, broadband IP response characteristics may be measured from the analysis of the decay curve form. As has been shown (Seigel et al., 1980), the spectral IP response parameters based on the Cole–Cole representation may be obtained directly from such analyses. In noisy areas, however, the frequency domain with its narrow-band filtering yields better S/N but less IP information per measurement.

A typical MIP system, suitable for making both time- and frequency-domain measurements, is shown in Fig. 12. It includes the following items:

Magnetometer (Scintrex MFM-3):

Fluxgate type
Sensitivity $100 \, \text{mV}/\gamma$
Frequency response: flat $0-1000 \, \text{Hz} \pm 3 \, \text{dB}$
Electronic noise level: $< 1 \, \text{m}\gamma \, \text{RMS}/\sqrt{\text{Hz}}$, $1-1000 \, \text{Hz}$.

Transmitter (Scintrex (TSQ-3):

Time and frequency domain
Maximum power: 3500 VA output
Maximum voltage: 1500
Maximum current: 10 A
Current stabilisation: $\pm 0.1\%$ for up to 20% external load variations or up to $\pm 10\%$ input voltage variation
Output waveforms: (a) Square wave, 0.1, 0.3, 1.0 and 3.0 Hz
                     (b) Time-domain interrupted square wave, 1,2,4 or 8 s (on/off times).

Receiver

(a) *Frequency domain* (Scintrex IPRF-2):
Measurements: PFE and RPS
Fundamental operating frequencies: 0.1, 0.3, 1.0 and 3.0 Hz
V_p Range: 100 mV to 10 V.

(b) *Time domain* (Scintrex IPR-10A):

Chargeability measurements: 1, 3 or 6 channels, switch selectable, of transient waveform, over current-off time of 1, 2, 4 or 8 s
V_p measurements: $30 \mu V$ to 30 V in 12 ranges
SP buckout: manual $\pm 1 \, V$
                automatic tracking to $20 \times V_p$
Signal enhancement: digital stacking of chargeabilities and V_p
                    continuous summing and averaging of chargeabilities and V_p.

FIG. 12. Typical time/frequency domain MIP survey system. 1, TSQ-3 transmitter; 2, MFM-3 magnetometer; 3, IPR-10A time domain receiver; 4, IPRF-2 frequency domain receiver.

6. FIELD RESULTS

In order to equip the reader better for an understanding of MIP field results, it will be useful to consider, at least qualitatively, some characteristic MIP response patterns. Figure 13 shows five typical MIP responses associated with vertically dipping strata. In type A we see a dike-like polarisable body which is more resistive than its neighbours. This could, for example, simulate a gold-bearing quartz vein, or a highly silicified zone carrying 2–5% disseminated sulphides. In type B we see the same case but with no resistivity contrast between the dike and its neighbours. In type C the dike is more conductive than its environment, which is a fairly common occurrence for a sulphide-bearing stratum. In type D the dike is more conductive than its environment but is non-polarisable, whereas the rocks on either side are polarisable. This is a very important case, simulating a massive sulphide body with disseminated sulphides in the wall rock on either side. It is well known that there is no volume-induced polarisation inside a well-interconnected, massive sulphide body. In this case the central massive core may act solely as a return current flow path for polarisation currents arising from the sulphide dissemination on its flanks.

The last example shown, type E, is one where the dike is both polarisable and more conducting than the strata on either side, but these strata differ in conductivity. The resulting responses show a concentration of return current flow on the more highly conducting flank, giving rise to an asymmetric response pattern (almost a sine wave). The interior current flow peak is shifted from above the body location to the point where the body may even be near the cross-over of the response curve.

7. CASE STUDIES

7.1. Spargoville, Western Australia

A small, sub-economic nickel sulphide body, about 250 m in length, buried under 30 m of oxidation, is located on the contact between an ultrabasic intrusive and an amphibolite rock unit. It is known, from surface resistivity measurements, that the fresh ultrabasics have an intrinsic resistivity of about 20 Ωm, whereas the fresh amphibolites are over 10-fold higher in resistivity.

A current electrode separation (2L) of about 360 m was employed, with the sulphide zone more or less in the middle of the centre line of the

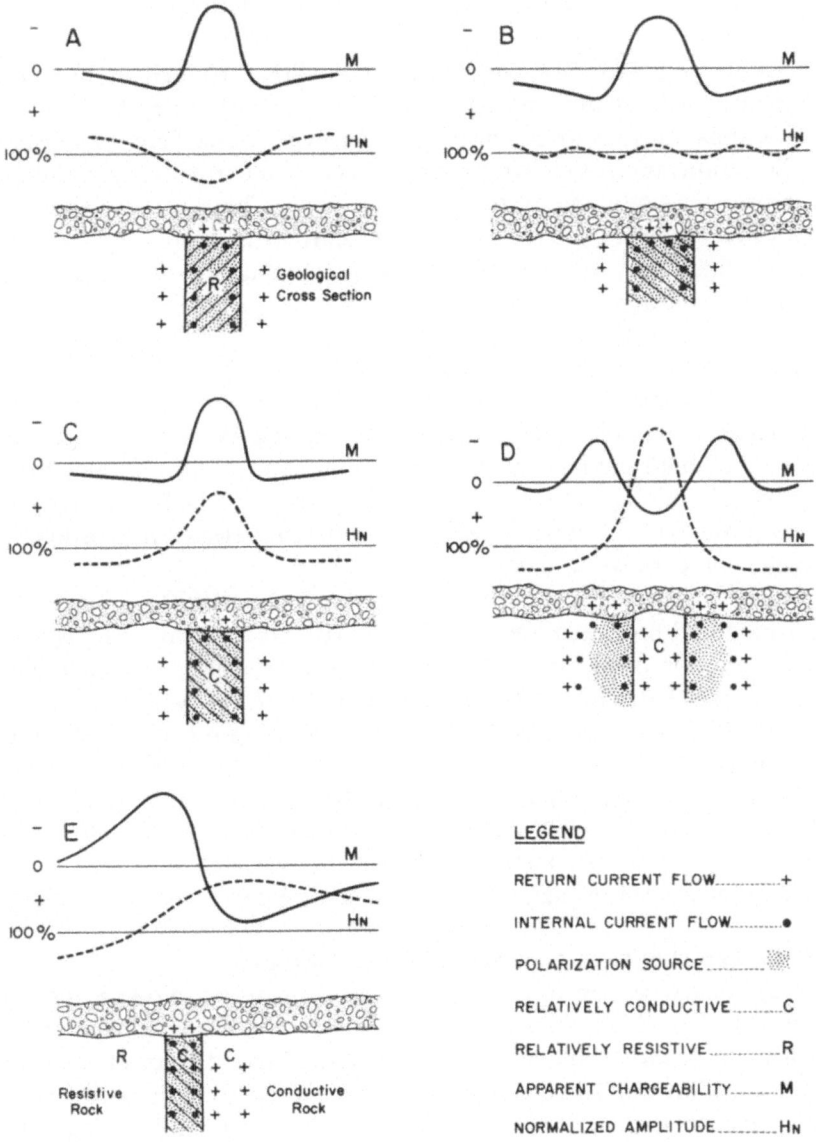

FIG. 13. MIP-type anomalies. A, IP with source relatively resistive; B, IP source with no resistivity contrast; C, IP source relatively conductive; D, IP source very conductive with disseminated halo; E, IP source on contact between resistive and conductive rock types.

spread. A 1 Hz square waveform was utilised. The results of the test measurements are shown in Fig. 14, in the form of H_N (as defined above), H_{SQ} (quadrature component) and ΔH_{SP} (change in in-phase component with change in frequency, derived from the PFE).

It is apparent that we are dealing with a type E case here, as attested by the various profiles shown. H_N shows increased current concentration, both over the body and on the ultrabasic flank. The H_{SQ} curve shows a sine wave response pattern, with interior current flow polarity over the amphibolites and return current flow predominating over the more conducting ultrabasic rocks.

7.2. Kalgoorlie, Western Australia

This case is interesting in that it concerns exploration for zones of disseminated sulphides, of possible gold-bearing potential, which are overlain by both salt lake material and tailing sand dumps resulting from nearby gold-mining activities. These tailing sands are highly conductive, of the order of $0.7\,\Omega\,m$ resistivity, thus rendering electrical prospecting methods inoperative to date.

Figure 15 presents the MIP. response on line 3650N. Quantities measured and presented are the RPS, MMR and H_{SQ}/I at 1 Hz. A very strong interior polarisation (type C) response has been indicated. The depth to the chargeable material was interpreted to be 70–80 m.

A vertical hole was drilled on the axis of this anomalous response. Disseminated and stringer pyrite, up to 10% locally, was encountered in black shales from 62 m depth. The top 47 m of the hole was composed of tailing sands, overburden and weathered rock. This example illustrates the ability of the MIP method to detect disseminated sulphide targets under almost 50 m of highly conducting surface cover.

7.3. Broken Hill Area, New South Wales, Australia

The geological section in this test consists of a steeply dipping zone of typical Broken Hill type lead–zinc mineralisation, 6 m thick, lying under 60 m of transported alluvial cover and an additional 30 m of oxidised rock. The alluvium is known, by means of electrical soundings, to have a resistivity ranging from 5 to $200\,\Omega\,m$. Borehole logs indicate that the fresh Precambrian rocks have resistivities in the 500–$10\,000\,\Omega\,m$ range and that the ore zone itself is not highly conducting, having a resistivity of about $100\,\Omega\,m$. It has a high (4:1) sphalerite/galena ratio, moderate pyrite content, and a coarse, crystalline texture.

Figure 16 shows the MIP results obtained over this rather deeply

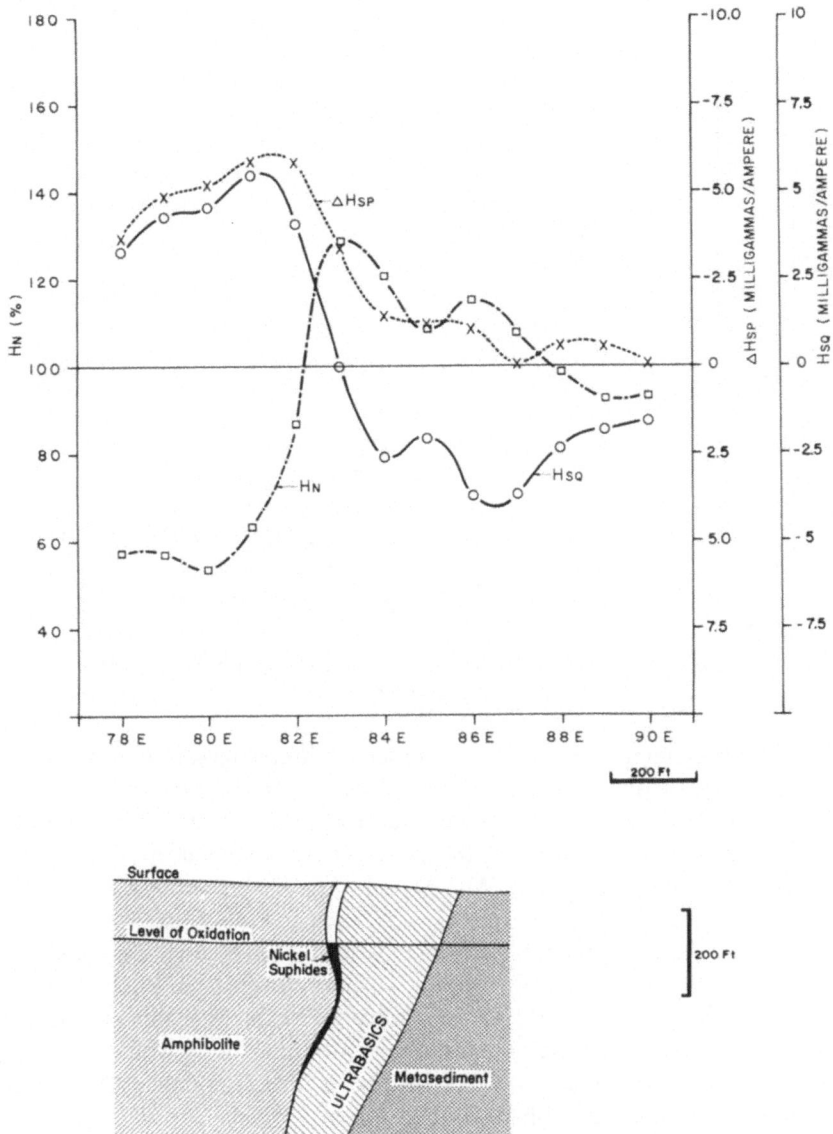

FIG. 14. Spargoville Deposit, Western Australia: frequency-domain MIP.
□—·—□, H_N, normalised amplitude; ○——○, H_{SQ}, quadrature component;
×-----×, ΔH_{SP}, amplitude change. (Courtesy: Selcast Exploration Ltd.)

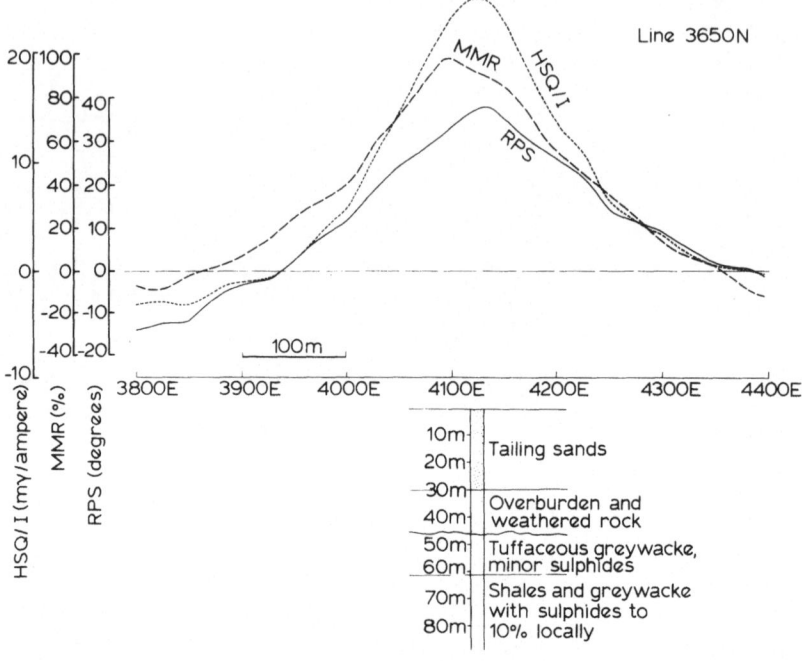

FIG. 15. Kalgoorlie, Western Australia: frequency-domain MIP. (Courtesy: N. L. Sanidine.)

buried target, using a 300 m electrode separation. Time-domain measurements were made using a 2 s on/off interrupted square wave current form. The Scintrex IPR-8 receiver was employed where chargeabilities corresponding to six slices of the decay curve, each 260 ms long, were measured. Three of these, viz. M_1, M_3 and M_5, have been plotted, as well as H_N, in Fig. 16.

It may be noted that this response pattern is of type C, characterised by a reasonable symmetric interior current flow M anomaly and an increased H_N over the body. One significance of this example is that it is possible for MIP measurements to detect a tabular body whose depth to width ratio is as great as 15:1, largely as a consequence of passing current longitudinally with respect to the body strike.

7.4. Widgiemooltha, Western Australia
A zone of massive nickeliferous sulphide mineralisation lies in a fold on the contact between Precambrian serpentinised ultrabasic rocks and amphibolites. Weathering extends to about 45 m and the resistivity of the

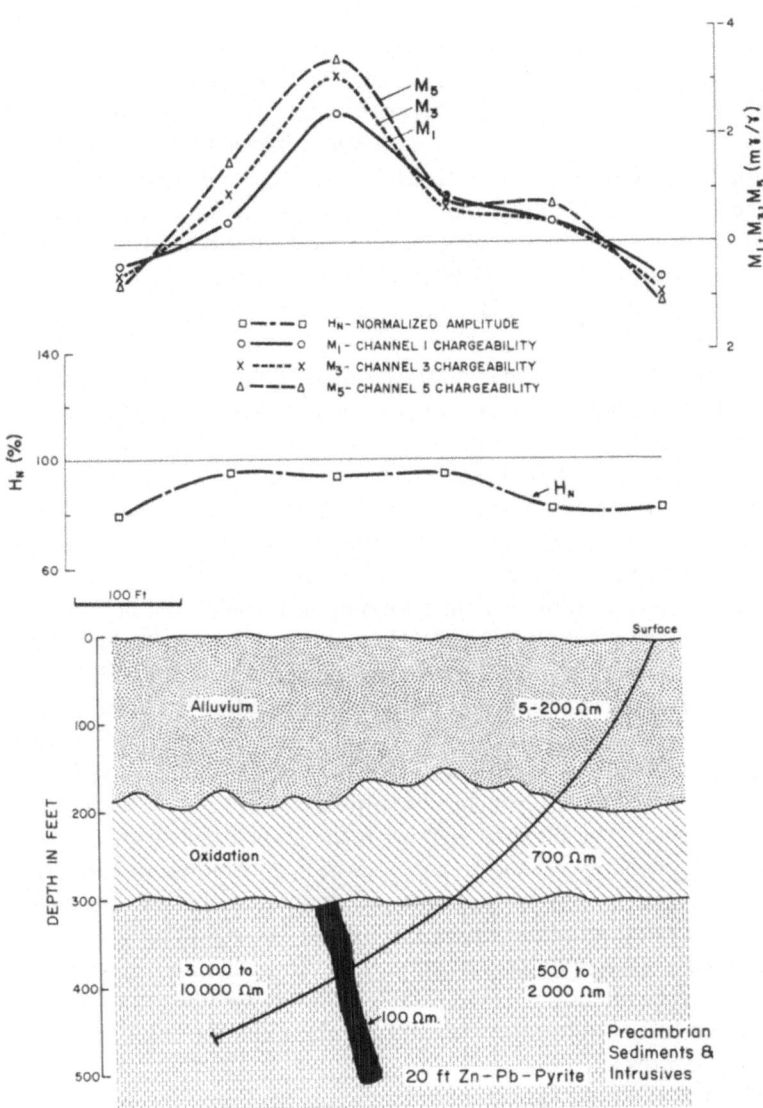

FIG. 16. Broken Hill area, Australia: time-domain MIP.

weathered zone has been measured to be between 3 and $10\,\Omega\,m$, thus providing a possible masking problem relative to the unweathered rocks below.

A 360 m longitudinal current electrode array was put out, with one current electrode placed 240 m north and the other 120 m south of station 2E. Comparative EIP results were obtained using a 60 m dipole–dipole array with $N = 1$–4. Both EIP and MIP measurements were made in the time domain, employing a 2 s on/off timing, with three slices (mode 2) of the transient decay curve being recorded. These slices are each of 520 ms duration and extend in total from 130 to 1690 ms.

Figure 17 shows the EIP and MIP results, as well as the geological section derived from the drilling, on line 4N. For both the EIP and MIP results the third slice (time 1170–1690 ms) of the transient response has been plotted. The ore zone in this section consists of two easterly-dipping lenses which would sub-outcrop at the base of the weathered zone, at about 1E and 2E. These lenses grade between 1 and 3% nickel and have a combined width of up to 12 m. The MIP results have been presented as H_N and H_{s3} (secondary field corresponding to the third slice, normalised with respect to the total current in the ground).

The composite sulphide zone shows up as a 168% conduction anomaly on the H_N curve, with dual peaks 45 m apart suggesting its dual character. The H_{s3} profile, on the other hand, indicates that the sulphide zones are acting as a return current path, thus providing an asymmetric type D response. The shape of the H_{s3} profile may be interpreted to yield a depth to the return current axis of about 45 m, which would intersect the upper edge of the nickeliferous sulphides.

There is an interior current flow in the serpentinised ultrabasic rocks west of 0, probably related to the polarisation response of the serpentine minerals as well as to some minor sulphide contribution. It is apparent that the massive sulphide lenses are not responding directly as IP sources but rather as preferential return current flow channels for polarisation sources in the ultrabasic rocks. The reason for this phenomenon has been suggested as the high resistivity contrast between the sulphide lenses and their host rocks.

The EIP results, shown in pseudo-section form, indicate a shallow zone of low resistivity over the sulphide lenses, but only on $N = 1$. The corresponding chargeabilities are very low in this area, suggesting a conducting, non-polarisable overburden response. The main EIP feature is a broad area of high chargeability lying over the ultrabasic side of its contact with the amphibolite, with the chargeability increasing pro-

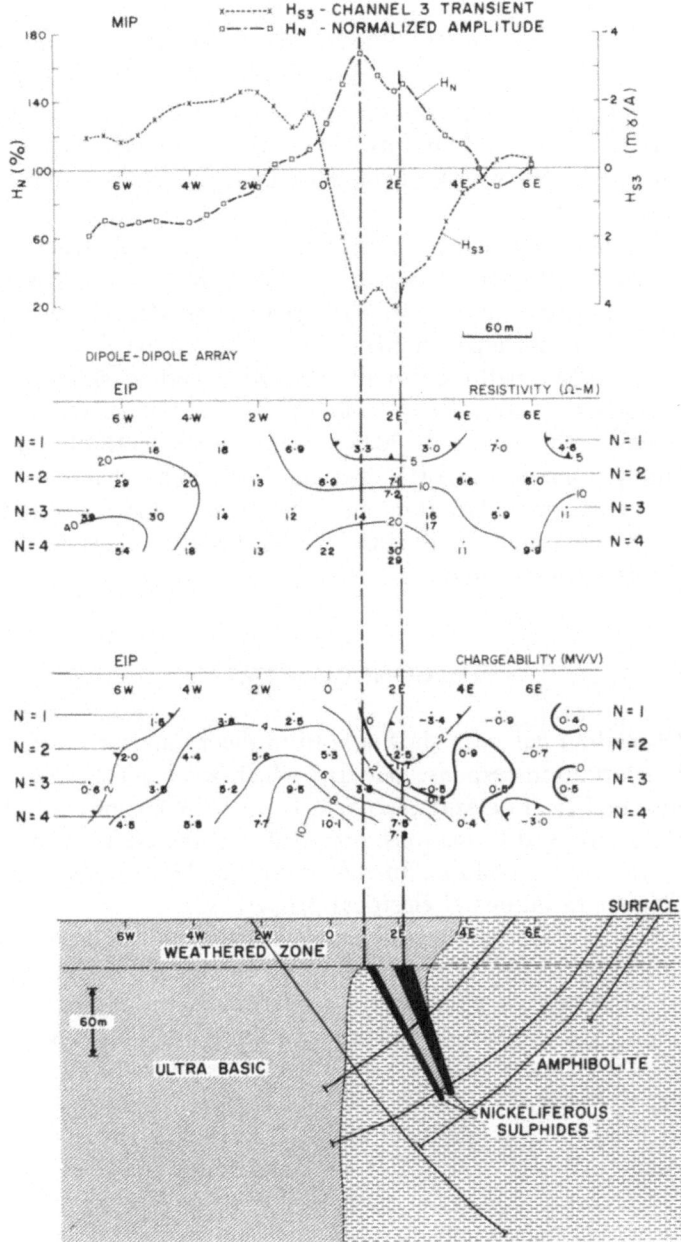

FIG. 17. Widgiemooltha Area, Western Australia: time-domain MIP and EIP. (Courtesy: International Nickel Australia Ltd and Broken Hill Pty Co. Ltd.)

gressively with depth (or N). Whereas the upward projection of the sulphide zone would fall under the eastern edge of this high chargeability area, the apparent resistivities in this region are on the rise. Conventional interpretation of the EIP results alone would probably have led to drilling for a target which might sub-outcrop at about 2W, clearly missing the ore zone.

Important factors which are illustrated by this example would include the following. Firstly, some 'massive' sulphide bodies may not respond as normal polarisation sources to IP surveys (EIP or MIP). However, such bodies are often indicated on MIP surveys as a type D anomaly (i.e. return current flow path). Secondly, the MIP method shows excellent spatial resolving power, being able to resolve the presence of two sulphide lenses which are only 30 m apart, under a depth of about 45 m of oxidation. Finally, the MIP method has hereby demonstrated its ability to provide good detection of sulphide bodies under a considerable thickness of moderately conducting overburden (conductivity–thickness product greater than 5 mho).

8. CONCLUSIONS

The MIP method has been shown, both in theory and practice, to yield induced polarisation response patterns which are significantly different from those of EIP. For example, there is no MIP response at all from horizontally stratified formations, provided that the electrical properties of each stratum are uniform. In that sense the MIP method responds only to lateral variations of electrical properties.

The MIP response from typical geological targets has been shown to exhibit a variety of patterns, depending on the local distribution of the resistivity and induced polarisation properties of the targets and their environment. On the basis of experience and computer models, these patterns may be interpreted to yield the more significant parameters of the causative bodies.

One of the particular merits of the MIP method is that it may detect the presence of polarisable materials under highly conducting (saline) overburden, such as occurs for example in parts of Australia. In such an environment EIP measurements often yield no useful information.

As theoretically predicted, MIP field results have confirmed that some highly conducting sulphide bodies may not yield an appreciable direct IP response. These bodies are sometimes detectable nevertheless by MIP measurements, by virtue of polarisable material on their margins, displaying what we call a characteristic type D MIP pattern. It may be possible to differentiate between zones of 'massive' and disseminated sulphide bodies through the difference in their MIP response patterns.

ACKNOWLEDGEMENTS

Many of the illustrations presented herein originally appeared in *Geophysics*. Permission to reproduce these has been generously granted by the Editor.

REFERENCES

EDWARDS, R. N. (1974) The magnetometric resistivity method and its application to the mapping of a fault. *Can. J. Earth Sci.* **11**, 1136–56.

EDWARDS, R. N. and HOWELL, E. C. (1976) A field test of the magnetometric resistivity method. *Geophysics* **41**, 1170–83.

HOWLAND-ROSE, A. W., LINFORD, G., PITCHER, D. H. and SEIGEL, H. O. (1980) Some recent magnetic induced polarization developments. Part I, Theory. *Geophysics* **45**, 37–54. Part II, Survey results. Ibid. **45**, 55–74.

SEIGEL, H. O. (1959) Mathematical formulation and type curves for induced polarization. *Geophysics* **24**, 547–65.

SEIGEL, H. O. (1971) Some comparative geophysical case histories of base metal discoveries. *Geoexploration* **9**, 81–97.

SEIGEL, H. O. (1974) The magnetic induced polarization method. *Geophysics* **39**, 321–39.

SEIGEL, H. O., EHRAT, R. and BRCIC, I. (1980) Microprocessor-based advances in time domain IP data collection, in-field processing and source discrimination. Presented at the 50th Annual Meeting of the Society of Exploration Geophysicists, November 1980.

SUMNER, J. S. (1979) The induced polarization method. In *Geophysics and Geochemistry in the Search for Metallic Ores*, Geol. Surv. Can. Econ. Geol. Rep. 31, pp. 123–33.

WAIT, J. R. (ed.) (1959) *Overvoltage Research and Geophysical Applications*, Pergamon Press, London.

Chapter 4

BROADBAND ELECTROMAGNETIC METHODS

J. W. Motter

Whitney & Whitney, Inc., Reno, Nevada, USA

SUMMARY

Recent applications of broadband electromagnetic methods used in exploration geology are covered in this chapter. Historically, electromagnetic detection systems used in mineral exploration efforts were designed to utilise a single frequency or a limited number of discrete frequencies. Application of electromagnetic detection methods in geologically complex environments has created a need for a more complete method of discerning the total geological environment. The objective is to be able to map the total geoelectric section, including resistive as well as conductive units. 'Broadband' electromagnetic methods have been developed to meet these needs. Electromagnetic detection systems capable of spanning up to five decades of spectral range are in current use. Recent developments in hardware technology and also in in-field data collection, processing, and sophisticated interpretational analysis are discussed in the chapter. These recent advances have contributed to our ability to resolve details of the three-dimensional distribution of resistivity in the subsurface. Results of both frequency-domain and time-domain surveys are presented.

1. INTRODUCTION

Electromagnetic (EM) prospecting systems employing either a single frequency or a limited number of discrete frequencies have been used in mineral exploration since the mid-1920s for detecting massive sulphide bodies or other targets possessing a conductivity contrast with the surrounding environment. Historically, many electromagnetic exploration systems have employed 'frequency-domain' equipment in the range of 100–5000 Hz and transmitter–receiver separations of between 30 and 300 m. Measurements of natural or artificial time-varying electromagnetic fields can yield considerable information concerning the distribution and nature of electrical parameters in the subsurface environment. Alternatively, 'time-domain' measurements may be made of the decaying fields in a conductor after the transmitter shutdown.

Recently, applied electromagnetic research has not only been used for locating metallic mineral deposits. Increasing use of EM methods has been made in exploration for coal, geothermal energy sources and permafrost (Vozoff, 1980). Electromagnetic surveys have also more recently been used to predict earthquakes, to map regional structural geology using both active source and natural source fields, and even to search for hydrocarbons in sedimentary basins (Duncan *et al.*, 1980). Exploration for these relatively resistive materials has encouraged the development and application of techniques using higher frequencies and utilising a much broader spectral range than had previously been employed. Moreover, broadband methods have arisen from the need to map the total geoelectric section, including conductive and resistive units, and recognising contributions of both local induction and current gathering involved. The focus of this chapter will be on broadband EM methods utilised in the search for metallic minerals, with which the author is most familiar. Although this chapter will focus on metallic mineral exploration, many of the concepts discussed herein are directly applicable to geothermal, coal and other targets.

It is not easy to define precisely what constitutes a truly 'broadband' electromagnetic system. The term 'broadband' is used to indicate those electromagnetic systems which, by virtue of a wide range of frequencies available and/or significant freedom in choice of available spacings and coil configurations, can excite all the components of the geological environment (e.g. massive and disseminated sulphides, host rocks, overburden, etc.). The frequency ranges employed in broadband electromagnetic systems vary widely; a typical range for commercially available units is from perhaps one to three decades of frequency. Research, development and prototype systems typically span even wider ranges. Ward *et al.* (1974*b*) describe the University of Utah's '14-frequency' system with a range of almost four decades (10.5 Hz to 86 kHz), while Duncan *et al.* (1980) describe a pseudo-noise-source very-wide-band system with a frequency band spanning over five decades (0.03 Hz to 15 kHz).

The purposes of this chapter are: to attempt to define the geological problems which necessitate the use of broadband EM methods; to survey equipment and methodology available for broadband data acquisition; and to show the importance of advanced interpretational and modelling techniques. Data from representative case histories will be referenced as examples of the manner in which interpretation of broadband EM survey data can significantly enhance our understanding of a complex geological environment.

2. THE PROBLEMS

In exploration for massive sulphide deposits or other conductive bodies using the active source ground electromagnetic method, an attempt is made to recognise a distinctive 'signature' of the body. The distinctiveness of each pattern is ultimately governed by the signal-to-noise ratio at the EM receiver. The amount of electromagnetic signal reaching a given depth in the earth is directly related to the earth's resistivity and inversely related to the frequency of electromagnetic source radiation. That is, the lower the frequency of the source radiation or the higher the earth resistivity, the greater the effective depth of penetration of the radiation. Effective depth of exploration depends largely on EM system gain, and on achievable signal-to-noise ratio.

Figure 1 (after Won, 1980) is a skin-depth nomogram relating source frequency, ground conductivity and depth of penetration. The skin depth is the depth in the earth where the field strength falls off to $1/e$ (approximately 37%) of the initial field strength. Indeed, the signal-to-noise ratio is the ultimate limit on the effective 'depth of exploration', the depth at which a target can be recognised as an anomaly. At least four components of noise can be defined in active source electromagnetic methods: (1) disturbance field noise, (2) instrument noise, (3) cultural noise and (4) geological noise.

Disturbance field noise consists of naturally occurring electromagnetic

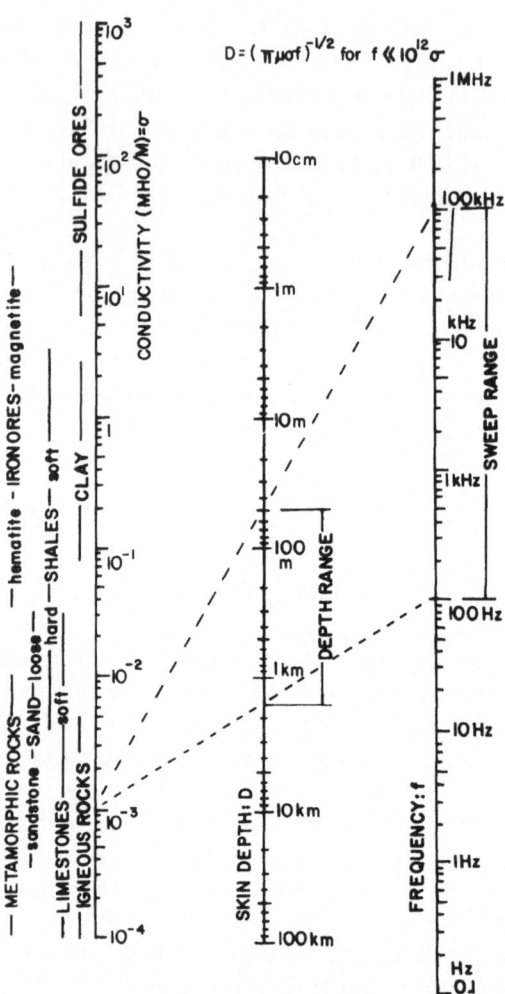

FIG. 1. Nomogram showing the relationship between source frequency, ground conductivity and depth of penetration (skin depth). A broadband or sweep-frequency EM with a frequency range from 100 Hz to 100 kHz would have a depth of exploration ranging from about 40 to 1500 m. (From Won, 1980.)

fields such as spherics, whistlers and numerous other sources. The fluctuations of the electromagnetic field in the frequency range being surveyed may be caused by thunderstorms or other sources. The problems caused by disturbance field noise can usually be overcome by increasing the strength of the active source transmitter. The weight of the required transmitters can often become the limiting factor. Natural electromagnetic field noise increases as frequency decreases; 10^{-1} Hz is a convenient lower cut-off frequency used in many portable induced polarisation (IP) systems currently on the market.

Instrument noise consists of both short-term variations as well as longer-term variations referred to as 'drift'. Modern solid-state technology used in current broadband instruments has reduced instrument noise to a level less than disturbance field noise and it is negligible for much survey work. Because of improvements in the electronic circuitry, the most significant contribution to instrument noise is noise mechanically induced in the detecting coils, particularly vibration of the coils, changes in effective area and magnetostrictive effects. With lower frequencies being employed in current and prototype broadband EM units, the effects of instrument noise will be significant in some cases.

Cultural noise consists of anomalies generated by the presence of man-made sources such as metallic buildings, pipelines, fences, power lines and telephones, and ground return from high-tension power distribution systems. This type of cultural noise may be significant in any particular survey and, in any event, must be dealt with on a case-by-case basic. Another component of cultural noise, sometimes classified as disturbance noise (Ward, 1967), is electrification noise. An example of this is electromagnetic radiation produced by power lines and telephone lines. Although electrification noise is a problem in some surveys, judicious selection of frequencies and filter circuitry in most broadband EM units largely eliminates this source of noise as a problem.

Geological noise is by far the most significant source of noise in broadband (or indeed in any) electromagnetic survey. Geological noise is defined herein as all anomalies arising from non-economic sources. Ward (1972) has described the problems of exploration for massive sulphides in a generalised geological environment as shown in Fig. 2. As illustrated in this figure, a realistic depiction of the geological environment in the vicinity of a base metal massive sulphide deposit may be very complex. Typically, one or more massive or concentrated sulphide deposits will be encased in a halo of variably disseminated sulphides. These disseminated sulphides may consist of discrete grains of variable size, or stringers, veinlets or veins, or some combination thereof. The halo of disseminated sulphides is surrounded by a host rock which may be bounded by an irregular top surface under a soil and/or rock overburden (Ward et al., 1974b). The problem is compounded in areas which have, or have had, tropical or arid weathering where the surface may be variably weathered

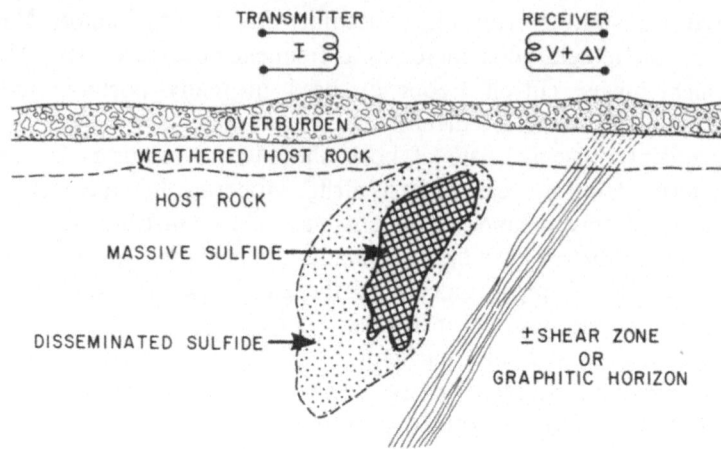

FIG. 2. Schematic representation of the general geological environment encountered in the search for a massive sulphide deposit. (After Ward *et al.*, 1974*b*.)

to depths in excess of 100 m and/or contain highly conductive pore fluids. Each component of the geological environment (massive sulphides, disseminated halo, host rock, weathered zone and overburden) may be electrically inhomogeneous and anisotropic in the x, y, and z directions. Graphitic or carbonaceous horizons and fault zones are often present near the deposits.

The separation of the responses due to the target from those due to geological noise (combined signatures of disseminated halo, conductive host rock and/or overburden, shears, faults, graphite, etc.) is not an easy task. An anomaly due to geological noise may extend over either a broader area or a more restricted area than the anomaly caused by an economic deposit of massive sulphides. The term 'longer spatial wavelength' is synonymous with a wider area, while the term 'shorter spatial wavelength' is used for a smaller area. Using this frequency-domain terminology leads to the concept of a continuous spectrum of spatial wavelengths for anomalies and noise.

An electromagnetic anomaly may be separated from geological noise if one or more of the following conditions are met:

1. The spatial spectral content of the anomaly is different from that of the geological noise.
2. The variation of anomaly with frequency is different from that of the geological noise.

3. The variation of anomaly response to transmitter–receiver separation or configurations is different from that of the geological noise.

In an attempt to achieve this separation, electromagnetic surveys commonly use a broad band of frequencies (or the equivalent time-domain delays), a wide range of transmitter–receiver separations and several transmitter–receiver configurations (Ward, 1979).

The current goals in interpretation of electromagnetic data are to be able (1) to explore deeper and (2) to resolve the various components of the geoelectric section. It is also necessary to separate the response of massive sulphides from the response of a halo of disseminated sulphides. If the various components of the geological section responded independently of each other, the interpretational problem would be considerably simplified. Interaction between the components can be very complicated and make interpretation very difficult. A primary electromagnetic field will cause currents to be generated in the massive sulphide deposit which is being sought, in the conductive overburden, in nearby shears, in the disseminated halo and in the host rock. Currents flowing through the host rock may be gathered into the massive sulphide deposit, enhancing the anomaly, but also rotating the phase and making the anomaly appear both deeper and of higher resistivity than it actually is. Advanced interpretational techniques are currently being developed which will enhance the ability of geophysicists to resolve complex, 'real earth' environments. These techniques use scaled physical models and numerical and analytical approaches to both the forward and inverse problems.

A number of papers have been written which examine the problem encountered in applying the electromagnetic method to a real earth environment containing massive sulphide combined with conductive overburden and host rocks, buried and/or surface topography, a disseminated sulphide halo and conductive faults, shears or intraformational conductors. Such papers include those by Lowrie and West (1965), Roy (1970), Sarma and Maru (1971), Gaur et al. (1972), Gaur and Verma (1973), Ward et al. (1974a,b, 1977), Hohmann (1975), Lamontagne (1975), Palacky (1975), Verma and Gaur (1975), Spies (1976), Lajoie and West (1976), Lajoie (1977), Lodha (1977), Braham et al. (1978) and Pridmore et al. (1979). Ward (1979) has compiled a list of the effects reported in those papers. Table 1 (adapted from Ward, 1979) summarises the most important effects of the presence of a complex geoelectric earth.

TABLE 1

SUMMARY OF EFFECTS OF EXTRANEOUS GEOELECTRIC FEATURES IN ELECTROMAGNETIC SEARCH FOR MASSIVE SULPHIDES (AFTER WARD, 1979)

Geoelectric feature	Effect	Interpretation problem
Conductive overburden	Rotates phase Decreases amplitude Decreases depth of exploration	Depth estimates invalid σt estimates invalid
Conductive host rock	Rotates phase Increases amplitude for shallow conductors Increases or decreases amplitude for deep conductors Changes shape of profiles Fall-off laws change	Depth estimates invalid σt estimates invalid Dip estimates invalid
Surface and buried topography	Introduces geological noise	Depth estimates invalid σt estimates invalid Dip estimates invalid May obscure sulphide anomalies
Disseminated sulphide halo	Rotates phase Increases amplitude	Depth estimates invalid Dip estimates invalid σt estimates invalid
Weathered host rock	Introduces geological noise	Obscures sulphide anomalies May invalidate all quantitative interpretation
Faults, shears, graphitic structures, other intraformational conductors	Introduces geological noise	Obscures sulphide anomalies May invalidate all quantitative interpretation

Note: σt = conductivity–thickness product of massive sulphide body.

3. DATA ACQUISITION

3.1. General

In order to be able to resolve a complex geoelectric earth into its components, it is necessary to be able to excite each of the components of the earth. Electromagnetic survey designs must produce excitation of both lateral and horizontal conductivity inhomogeneities. The operational choices in designing a broadband survey are as follows:

1. Should electromagnetic profiling or sounding or a combination sounding–profiling technique be employed?
2. Should time-domain or frequency-domain measurements be made?
3. What transmitter–receiver configuration should be used? What spacings or range of spacings should be utilised?
4. What frequency range should be used and what geoelectric parameters should be measured?

Obviously a number of the questions are interrelated, and the choice of equipment selected for a particular survey will depend on the survey environment and the nature of the conductive target searched for.

3.2. Sound-Profiling Measurements

'Profiling' is done for the purpose of locating lateral conductivity inhomogeneities in the near-surface portion of the earth's crust. The transmitter in an active source electromagnetic profiling system will typically consist of a single- or multi-coil loop of wire erected in either a horizontal plane (e.g. vertical magnetic dipole source, VMD) or vertical plane (e.g. horizontal magnetic dipole source, HMD) and energised with a flowing current. The transmitter may be stationary while measurements of the magnetic field components are made, or the transmitter and receiver may be moved together.

Components of the magnetic field which may typically be measured in a broadband electromagnetic survey are the orthogonal cartesian components of the field H_x, H_y and H_z or, equivalently, the descriptors of the ellipse of polarisation α and ε, the tilt angle and ellipticity respectively. Figure 3 illustrates the descriptors of the ellipse. The tilt angle, α, is the inclination from the horizontal of the major axis of the ellipse. The ellipticity, ε, of the ellipse is the ratio of the minor to the major axes of the ellipse. The behaviour of α and ε across a hypothetical conductor are shown in Fig. 4 (after Pridmore, personal communication, 1978). Any of the above five field parameters can vary systematically in the vicinity of a

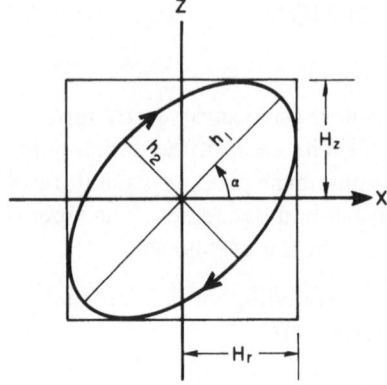

Fig. 3a. Geometric representation of ellipse of polarisation. H_r is the real part of the total magnetic field, H_z is the imaginary part of the total magnetic field. The magnetic fields are $H_r \exp i\phi r$ and $H_z \exp i\phi z$ with phase difference $\Delta\phi = \phi_z - \phi_r$.

$$\text{Tilt angle: } \tan 2\alpha = \frac{2 \left| \dfrac{H_z}{H_r} \right|}{1 - \left| \dfrac{H_z}{H_r} \right|^2} \cos(\phi_z - \phi_r)$$

$$\text{Ellipticity} = \pm \frac{|h_2|}{|h_1|} = \varepsilon = \left| \frac{H_z \cos\alpha - H_r \sin\alpha}{H_z \sin\alpha + H_r \cos\alpha} \right| \text{sgn}(\phi_z - \phi_r)$$

Fig. 3(b). Descriptors of ellipse polarisation. The sign of ellipticity is dependent on the sign of $\Delta\phi$.

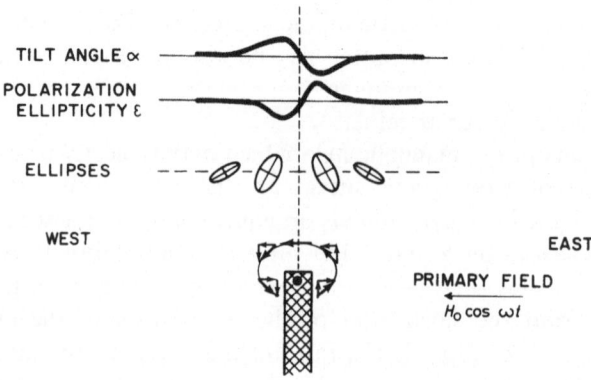

Fig. 4. Behaviour of α and ε across a hypothetical conductor ($90° < \Delta\phi < 180°$). Transmitter is an HMD source (vertical loop) east of the conductor.

conductor. Measurements of either α and ε, or H_x, H_y and H_z as functions of frequency and spatial (x, y, z) position yield information as to the depth, length, conductivity and other parameters of a subsurface conductive inhomogeneity. The major and minor axes of the ellipse of polarisation may assume any ratio or orientation in the vicinity of an inhomogeneity; most often, however, measurements of α and ε are made in the vertical plane passing through either a vertical or horizontal transmitting loop (Ward *et al.*, 1974*a*). Figure 5 illustrates the measurement of α and ε for the horizontal loop EM method.

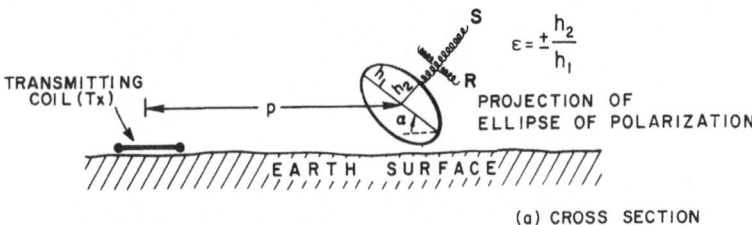

(a) CROSS SECTION

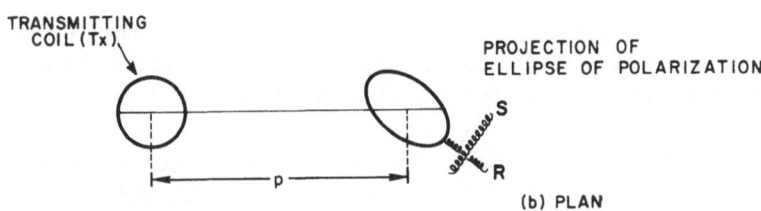

(b) PLAN

FIG. 5. Dual-coil method of measuring tilt angle, α, and ellipticity, ε, near a horizontal transmitter coil. S is the signal coil, R is the reference coil, p is the transmitter–receiver separation. (From Ward *et al.*, 1974*b*.)

The parameters to be measured in profiling depend on the transmitter–receiver configurations used. In the vertical loop method, the VLF method and the Crone shootback method, the tilt angle is measured. Surveys using horizontal loop methods for profiling typically measure the real and imaginary parts of the vertical magnetic field. In the Turam method, at least three types of measurements may be taken, depending on the instrumentation chosen. The first method measures the ratio $\text{Re}(H_{z1})/\text{Re}(H_{z2})$ of the real parts and the ratio $\text{Im}(H_{z1})/\text{Im}(H_{z2})$ of the imaginary parts of the vertical magnetic field, observed at locations 1 and 2, approximately 30 m apart. Figure 6 shows the operational set-up

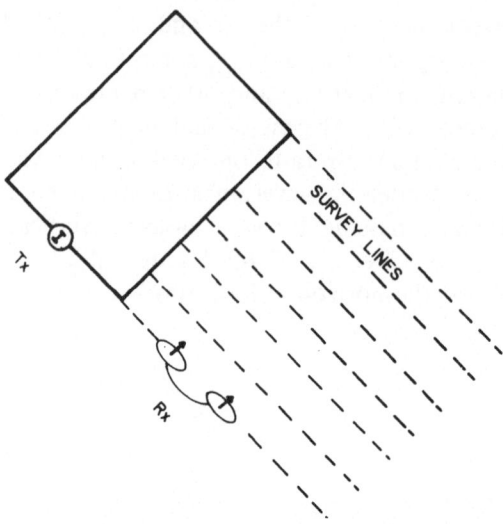

FIG. 6. Schematic illustration of the Turam method. Field strength/amplitude ratio and phase difference between the receiver coils are typically measured.

for a typical Turam-type survey, in which the phase difference $\phi(H_{z1}) - \phi(H_{z2})$ is measured. A third method, used in the Elfast EM (Androtex), can measure absolute amplitude $A_1(H_{z1})$ and $A_2(H_{z2})$ as well as the phase difference.

Electromagnetic sounding is a process whereby a horizontal loop or vertical loop transmitter (VMD or HMD source, respectively) may be used to determine the conductivities and thicknesses of layers within a horizontally stratified earth. If the frequency of the source is varied, 'parametric sounding' is conducted. If the transmitter–receiver separation is changed, 'geometric sounding' may be performed. It is not uncommon to combine the two sounding methods in order to trace out the variation of H_z, H_r, α or ε with induction number $B = (\sigma\mu\omega)^{1/2} r$ from a low to a high asymptote. In this equation, σ is the apparent conductivity of the geoelectric section, μ is the magnetic permeability. As discussed by Ward et al. (1974a) and Ward (1979), it may be necessary to vary the source frequency, ω, over four to five decades using parametric sounding, or the transmitter–receiver separation, r, over at least two to three decades to ensure that both high and low response asymptotes are achieved.

Vanyan (1967) describes component sounding techniques, while Dey and Ward (1970), Ryu et al. (1970), Ward et al. (1974a,b, 1976), Glenn and Ward (1976) and Dey and Morrison (1977) discuss tilt angle and ellipticity sounding techniques. Raiche and Spies (1981) discuss modelling of time-domain sounding and present master curves for the coincident loop case.

In general, horizontal loop EM sounding may be less useful than vertical loop methods for determining the structure of a horizontally layered earth. The total electric field for a horizontal transmitting loop is horizontal everywhere and may be represented by the tangential electric component E_θ which is complex, that is, phase shifted with respect to the loop current. The amplitude and phase of E_θ are determined by the induction number, B, and the geoelectric parameters of the layers, such as thickness and conductivity. Since the electric field does not cross horizontal boundaries in a layered earth, the horizontal loop EM sounding method is sensitive to conductive layers but much less sensitive to resistive layers. Since H_r and H_z have the same frequency, ω, but typically different phases, elliptical polarisation of the magnetic field vector results.

EM sounding methods using a vertical loop transmitter, in contrast to the horizontal loop transmitter, utilise a field consisting of both E_r and E_z components. The electric field thus both crosses and parallels horizontal formation boundaries in the subsurface, and allows the vertical loop sounding to be sensitive to both resistive and conductive layers. Measurements of the magnetic field components H_r and H_z are made at locations along the axis of a vertical transmitting loop for electromagnetic sounding. Alternatively, α and ε may be measured in a vertical plane containing the axis of the loop.

Electromagnetic profiling and sounding may be combined in a single survey. Conventional Turam, for example, can be used for geometric sounding of a horizontally layered earth. If the frequency range of a Turam-type survey is expanded to four or five decades (as is possible with currently available equipment discussed subsequently), it may be used for parametric sounding as well. Utilising a generalised Turam system in both parametric and geometric modes simultaneously, it is possible to achieve continuous sounding–profiling.

A vertical loop transmitter may similarly be used in continuous profiling–sounding. The tangential magnetic field H_θ can be measured in the plane of the loop or the parameters H_r, H_z or α and ε can be measured along the axis of the loop for this purpose. Pridmore et al. (1979) present results of sounding–profiling over a massive sulphide deposit in California.

Profiling and sounding generate unique data sets which complement

one another. A combination of the two techniques allows more information concerning the geoelectric nature of the earth to be determined
than is available by either technique alone, regardless of the specific
purpose of the EM survey. The additional time and expense required to
collect broadband, continuous profiling–sounding information must still
be determined on a case-by-case basis. It is clear that the further
development and application of microprocessor-based broadband EM
units and automated recording of digital signals on magnetic media,
together with computer processing of the data, will enhance the ability to
collect profiling–sounding data on a cost-effective basis. It appears,
particularly in areas of complex geology, that the additional effort
required to generate, collect, process, and model broadband EM results
will be economically worthwhile.

3.3. Domain of Acquisition

Electromagnetic data are always collected as a time series describing one
or more parameters of an electromagnetic field at some point in time/
position/space, e.g. as a function of (x, y, z, t). Resulting data may be
processed and interpreted in either the frequency domain or in the time
domain. In the frequency domain the spectrum is viewed through a
frequency window or passband. In frequency-domain measurements, a
continuous sinusoidal (or other shape) waveform is applied to the
transmitter and the receiver is activated throughout the transmitting
interval.

In the time domain the transient decay (amplitude) of an impulsive
waveform is measured by the receiver through a time window or
passband following the cessation of current in the transmitter (Ward,
1979). In broadband frequency-domain electromagnetic surveys, obser-
vations of magnetic field parameters H_x, H_y, H_z or α and ε are made over
a range of one to five decades of frequency. At least four observations per
frequency decade are usually necessary to define adequately the shape of
the parameter response–frequency curve.

In broadband time-domain electromagnetic surveys, measurements of
the amplitude (or of the orthogonal components H_x, H_y and H_z of the
amplitude) are made at multiple time delays following current shut-off.
Alternatively, one or more of the three spatial components of the first
time derivative of the secondary magnetic field may be measured as a
function of position (McNeill, 1980). Measurements are typically made
with time delays ranging from 0.08 to 180 ms after current shut-off of the
primary field. The initial transient decay varies rapidly with time,

particularly in environments possessing a small time relaxation constant, τ. For this reason, the time windows or passbands employed must necessarily be very narrow to avoid distortion of the signal resulting from changes in signal amplitude during the sampling period. McNeill (1980), in a review of transient EM techniques, notes that by nature transient systems are broadband. As noted therein:

'A Fourier analysis of the emf induced in the target shows that odd-harmonic components exist from the basic pulse repetition-frequency (typically from 3 to 30 Hz) up to a value determined by the duration of the transmitter turn-off time; for example a 200 μsec turn-off puts significant excitation to frequencies of many kHz. This multi-spectral excitation is very desirable for obtaining detailed information about the target spectral response, however an accompanying disadvantage is that the receiver, being broadband, is susceptible to external noise and interference. In fixed-frequency systems, where the primary field is carefully cancelled out, the system noise is usually determined by errors in this cancellation and a point is soon reached where increases in transmitter dipole moment do not increase the signal-to-noise ratio. The situation is quite different for transient systems in which measurement is made during the transmitter off-time; for such systems the effective exploration depth is often set by external noise and the availability of a large transmitter loop with a high value of transmitter current becomes of great importance.'

Time-domain measurements are related to frequency-domain measurements through the Fourier transform. Despite considerable debate the two instrumental approaches are *theoretically* equivalent and there is no particular reason why one transmitted waveform cannot be processed and interpreted in either the frequency domain or the time domain or both domains simultaneously (Ward, 1979). McNeill (1980) has suggested, however, that while the two domains are theoretically equivalent, differences in the sources of noise for the two techniques lead to practical differences in the relative merits of the two instrumental approaches, often favouring transient techniques. A detailed discussion of his arguments is beyond the scope of this chapter, but in essence his reasoning is based on the fact that the transient method allows measurement of in-phase components of the response in the absence of a primary field, as opposed to frequency-domain measurements which are made with the primary field on. McNeill notes that the frequency-domain measurements may be limited by inability to remove the primary field from the resultant field and by inaccuracy of the transmitter–receiver separation and the fact that both signal and noise increase linearly with transmitter power, while with transient techniques the signal-to-noise ratio almost invariably increases linearly with transmitter dipole moment. A powerful transmitting source is thus very desirable in a pulse or transient system. A disadvantage of transient systems is that, since with a transient EM system resonant tuning techniques are not available, the transmitters tend to be large and heavy and use large areas to achieve high moments. For this reason, current transient EM systems are not as portable as many frequency-domain systems.

The relative merits of time-domain versus frequency-domain broadband EM systems is thus a subject of ongoing debate and it appears that the advantage that one or the other may possess in any given circumstances is due entirely to the advantages or limitations of comparative instrumentations.

Optimum waveform design is an area of considerable recent research. The UTEM system (Lamontagne and West, 1973) was designed to transmit a triangular waveform and to receive its derivative, a square waveform. The EM Surveys, Inc., transmitter is microprocessor-controlled, and is capable of transmitting waveforms of any specified shape by specifying a suitable pattern of digital amplitudes (Jeppsen, personal communication, 1981).

Pseudo-random binary sequence (PRBS) generators (Quincy et al., 1976; Duncan et al., 1980) and swept-frequency sources (Won, 1980; see also Chapter 2, this volume) have been proposed as composite waveforms to enhance signal-to-noise ratios.

An additional potential advantage of swept-frequency or PRBS pseudo-noise-source waveforms is the ability to collect data rapidly. Use of multi-frequency broadband frequency-domain electromagnetic systems requires making measurements at each of many frequencies. Thus, without introducing fast frequency-multiplexing or power-switching techniques, such a system could not be employed in a mobile (vehicular or airborne) installation (Won, 1980).

Figure 7 (from Duncan et al., 1980) is an example of a pseudo-random binary sequence used in a broadband EM system. A PRBS is characterised by two adjustable parameters: f_c, the clock frequency, and n, the sequence length parameter. The sequence repeats after $2^n - 1$ clock pulses. In Fig. 7, n is 5 and the PRBS repeats periodically after 31 clock pulses. The power spectrum of this signal is shown in Fig. 8. The line spectrum is equispaced in frequency by $f_c/(2^n - 1)$.

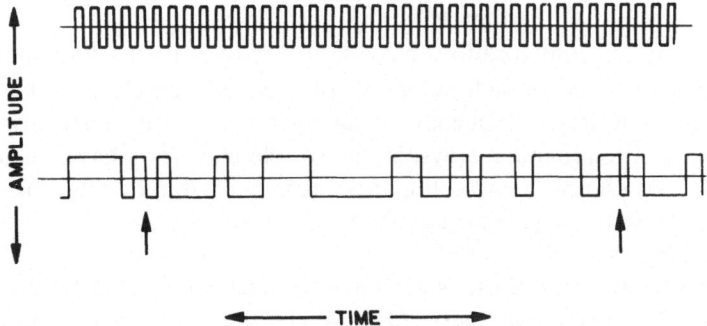

FIG. 7. Example of a pseudo-random signal generated using a shift register with programmed feedback from the clock signal above. The signal repeats after $2^5 - 1$ clock pulses and the arrows indicate the length of one period. (From Duncan *et al.*, 1980.)

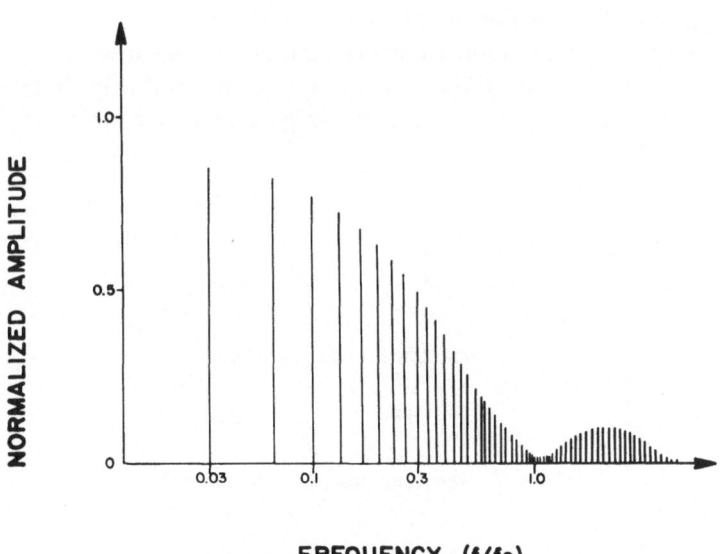

FIG. 8. The power spectrum of the pseudo-random signal shown in Fig. 7 plotted as a ratio of frequency, f, relative to clock frequency, f_c. (From Duncan *et al.*, 1980).

3.4. Choice of Coil Configurations

Numerous coil configurations have been employed in broadband electromagnetic surveys. In fact, nearly all the coil configurations which have
been used in single frequency or narrowband surveys have also been
applied to broadband surveys. Standard reference texts have treated the
variety and choice of coil configurations in substantial depth; the details
are not repeated here. Particularly good reviews are given by Grant and
West (1965) and Ward (1967, 1979).

Three main classifications of configurations were suggested by Ward
(1979): (1) roving coil pairs, fixed orientations; (2) roving coil pairs,
rotatable orientations; and (3) fixed transmitter, roving receiver. The
transmitting source employed in each case is a loop of wire energised by
a current. A loop may be oriented with its axis vertical (VMD or vertical
magnetic dipole source), termed a 'horizontal loop' source, or the loop
may be erected vertically with its axis horizontal. This latter source is
referred to as a 'vertical loop' source and has a horizontal magnetic
dipole (HMD). With the vertical loop source, measurements may be
made either along the axis of the loop or within the plane of the loop.
Similarly, the receiving coil in a fixed orientation survey may be oriented
along any of three orthogonal axes.

The three most common fixed orientations are illustrated in Fig. 9.
The first of these is termed the 'horizontal loop' method (Fig. 9(a)), where
both transmitter and receiver use VMD geometry and are transported

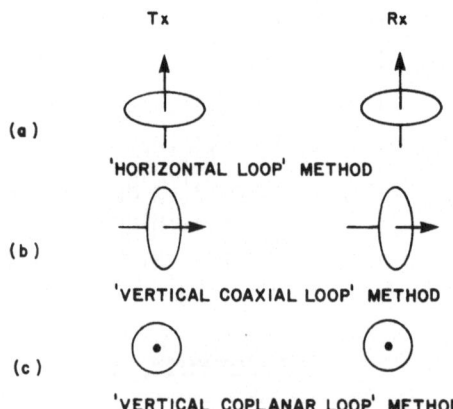

FIG. 9. Three common coil configurations using fixed transmitter–receiver
geometrics. Dipole moments are indicated by arrows.

with constant coil separator, in-line in a direction normal to the geological strike. Commercially available equipment which generally utilises this configuration is listed in Table 2 (updated after Hood, 1977, and Ward, 1979). (Some of this equipment may also be utilised in other configurations.) The two remaining fixed orientations are the 'vertical coaxial' (Fig. 9(b)) and 'vertical coplanar' (Fig. 9(c)) configurations. In surveys using the vertical coaxial method, both the transmitting and receiving coils use HMD geometry with their axes in line ('coaxial') and are transported with constant separation. The coil pair may be moved either in-line perpendicular to the geological strike or broadside in a direction perpendicular to the geological strike. The vertical coaxial configuration is the most generally used of the vertical loop configurations. Table 3 (after Hood, 1977, and Ward, 1979) lists some of the commercially available vertical loop electromagnetic units. It can be noted that none of these are truly broadband, spanning at best just over one decade of frequency.

The second major classification of coil configurations consists of roving coil pairs with rotatable orientations. The Crone shootback method falls in this class, and is notable in that the transmitting and receiving coils are interchangeable and that each is used as both a transmitter and a receiver. Utilising this strategy eliminates the effects of vertical elevation differences between transmitter and receiver. Details are given by Crone (1966, 1973).

Most of the truly broadband electromagnetic systems have been classified as fixed transmitter, roving receiver. Table 4 (modified after Hood, 1977, and Ward, 1979) lists commercially available 'fixed transmitter' electromagnetic systems. Some of these systems can be utilised with moving transmitter configurations, but typically are not, owing to transmitter weight or other operational considerations.

Four different coil configurations employing a fixed transmitter are commonly used for broadband electromagnetic surveys. The Turam method has been discussed previously and is illustrated in Fig. 6. The three other coil configurations are the vertical axial coil, the rotating vertical loop and the fixed horizontal loop configurations. Figure 10 illustrates the application of each of these three techniques over a massive sulphide body (from Pridmore et al., 1979). The orientation of the projection of the ellipse of polarisation on a vertical plane through the axis of the transmitter is shown in Fig. 11. Details of the survey methodology and interpretation of results are discussed in Ward et al. (1974a,b) and Pridmore et al. (1979).

TABLE 2

COMMERCIALLY AVAILABLE HORIZONTAL LOOP GROUND ELECTROMAGNETIC EQUIPMENT (AFTER HOOD, 1977, AND WARD, 1979)

Manufacturer	Model designation	Frequency of operation (Hz)	Coil separation (m)	Dipole moment in amp turns (m^2) (at spec. freq.)	Component measured (I/P, in phase; O/P, out of phase)	Readout device	Range of readings	Read-ability	Weight (kg)	Power source
ABEM	Demigun	880, 2640	30, 60, 90, 150, 180	50/880 Hz 20/2640 Hz	I/P and O/P	2 dials	0–160% I/P ±80% O/P	±0.5%	23.2	D or Nicad cells
Apex Parametrics	Max-Min II	222, 444, 888, 1777, 3555	25, 50, 100, 150, 200, 250	220/222 Hz 200/120/888 Hz 60/1777 Hz 30/3555 Hz	I/P and O/P	2 dials +tilt-meter	±100% I/P ±100% O/P	±0.5%	Tx: 13 Rx: 6	12V gel cells
	Max-Min III	111, 222, 444, 888, 1777	50, 100, 150, 200, 250, 300	600/111 Hz 550/222 Hz 300/444 Hz 150/888 Hz 75/1777 Hz	I/P and O/P	2 dials +tilt-meter	±100% I/P ±100% O/P	±0.25 to ±0.5%	Tx: 26 Rx: 6	24V gel cells
Geonics	EM17	1600	30, 60, 90, 120	24	I/P and O/P	Meter (self-indicating)	±100% I/P	0.5%	12.61	Rx: C cells
	EM 17L	817	50, 100, 150, 200	24 (reduced) 48(normal)			±50% O/P	0.25%	13.4	Tx: D cells
McPhar	VHEM	600, 2400	30, 60, 90 or 40, 80	60/600 Hz 18/2400 Hz	I/P and O/P	Dial and headset	±100% I/P ±100% O/P	±0.5%	8.2	Rx: 9V Tx: D cells
Scintrex	SE-600	1600	60 or 90	27	I/P and O/P	Dial and headset	±100% I/P ±50% O/P	±0.5%	15	6 and 13.5V cells

TABLE 3
COMMERCIALLY AVAILABLE DIP-ANGLE GROUND ELECTROMAGNETIC EQUIPMENT (AFTER HOOD, 1977, AND WARD, 1979)

Manufacturer	Model designation	Frequency of operation (Hz)	Maximum coil separation (m)	Dipole moment in amp turns (m^2) and weight (kg)	Transmitter power source	Weight of receiver (kg)	Readability of clinometer	Bandwidth of receiver system	Remarks
Crone Geophysics	CEM (shootback)	390, 1 830, 5 010	200	45/390 Hz 30/1 830 Hz 18/5 010 Hz 10 kg with batteries	3 × 6 V lantern batteries	11	±0.5°	5 Hz/390 Hz 15 Hz/1 830 Hz 30 Hz/5 010 Hz	Transceiver units for shootback and vertical loop
	VEM	390, 1 830	700	1 100/390 Hz 900/1 830 Hz 20 kg with battery	12 V/ 24 A h gel battery	6	±0.5°	5 Hz/390 Hz 15 Hz/1 830 Hz	High-powered vertical loop
McPhar	REM	1 000, 5 000	200	60/1 000 Hz 15/5 000 Hz 4.5 kg	Hg cells	2.4	±0.5°	20 Hz/1 000 Hz	
	VHEM	600, 2 400	200	60/600 Hz 18/2 400 Hz 4.1 kg	300 cells	3.8	±0.5°	13 Hz/600 Hz	Horizontal loop also
Scintrex	SE-600	1 600	300	27 8 kg	2 × 6 V cells	5.5	±0.5°	10 Hz?	Horizontal loop also

TABLE 4

COMMERCIALLY AVAILABLE FIXED HORIZONTAL LOOP GROUND ELECTROMAGNETIC EQUIPMENT (MODIFIED FROM HOOD, 1977, AND WARD, 1979)

Domain	Manufacturer	Model designation	Parameters measured	Frequency of operation (Hz)	Power output	Power source (MG = motor generator)	Readout device	Weight (kg)	Remarks
Time	Crone Geophysics	Pulse EM (PEM)	8 samples of secondary field, 1 sample of ramp voltage	Equivalent 18–1060 Hz	Max. 450 W	2 × 12 V 20 A h Bel batteries	Meter	21	Moving Tx-loop system
	Geonics	UTEM	Vertical magnetic and horizontal electric fields	10 time slots, base frequency 7–90 Hz		1.5 kW MG	Meter or digital tape deck	60	Large loop
		EM-37	20 samples of secondary field, 12.5–12 500 Hz	Max. 2800 W	5 h.p.	MG	LED display	Tx:20 Rx:30	Integration
	Geoex.	SIROTEM	Vertical magnetic field	12–32 time slots, 0.25–180 ms	Max. 176 W	22V/10A h Nicads	Printer output	20	Single or double loop configuration, moving loops, choice of sizes

Frequency							
Geoprobe	Maxi-probe EM 16	Vertical and horizontal magnetic field, horizontal electric fields	$2^n (n=0-9)$, 2, 4.1. 8.2, 16.4, 32.8, 41 kHz	2100 W 6 h.p. MG	Meter	116	
Scintrex	SE-77/ TSQ-2M (Turam)	Field strength ratio and phase difference	35, 105, 315, 945, 2 835	500 W 3 h.p. MG	Automatic meter display	42	Reads harmonics of transmitted square wave
Geotronics	EMR-1/ GT20A	32f	10–10000 Hz	20 kW MG	Meter	Rx:10	Coherent super without carried reference
McPhar	GEM-8	Phase and amplitude (converted to ellipse of polarisation parameters)	41, 82, 164, 328. 656 Hz, 1.312, 2.624, 5.248 kHz	500– 2 000 W MG	LED phase $\pm 0.1°$, amplitude $\pm 0.1\%$	Tx:11 Rx:10	Fixed loops to 30 m, portable 1 m loop, HMD or VMD Tx configurations
Androtex	ELFAST	Amplitude (V_1 and V_2) in Rx coils, phase	25, 75, 225, 675, 2025	5 h.p. MG	LED/LED	Tx:7.5 Rx:4.6	$Rx_1 - Rx_2$ separation typically 50–200 m
EM Surveys	DSP-2 MFD-3	Amplitude, phase	0.001–10 kHz	3 200 W MG	Digital output printer or magnetic tape	Rx:16 Tx:30	Microprocessor-controlled transmitters, controller and digital signal processor

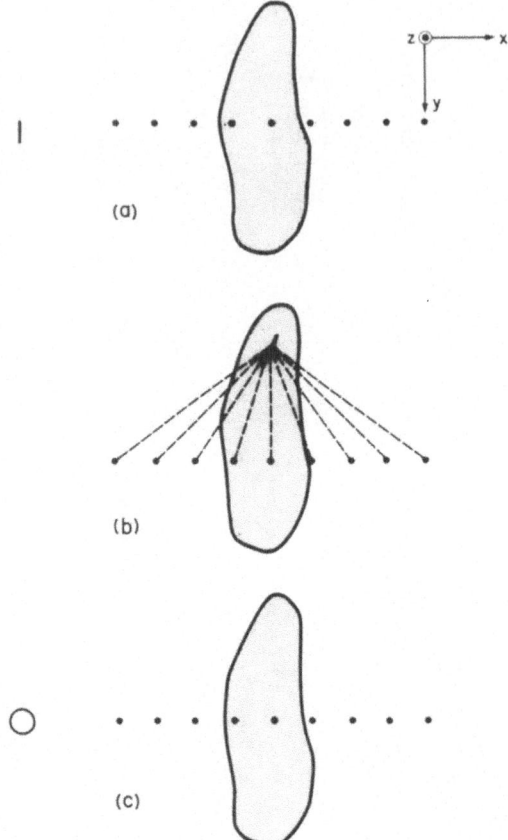

FIG. 10. Three coil configurations employing a fixed transmitter: (a) vertical axial loop; (b) vertical rotating loop; (c) horizontal loop. The shaded outline represents a massive sulphide body, the dots are the receiver locations.

FIG. 11. Vertical axial loop method utilising a fixed transmitter. The orientation of the ellipse of polarisation on a vertical plane passing through the axis of the transmitter is shown.

The Australian Sirotem time-domain electromagnetic system is unique in that the same coil functions as both the transmitter and the receiver. A single rectangular loop 50–200 m on a side is used first as a transmitter and then at appropriate time intervals as a receiver. Alternatively, the Sirotem can use separate transmitter and receiver loops.

An important parameter in survey design is the transmitter loop size used. It can be argued that the larger the loop size, the greater the depth of exploration for two reasons. Firstly, a small loop behaves like a dipolar source with a fall-off rate of $1/r^3$ whereas a large loop behaves more like four line sources with a fall-off rate of $1/r$. Secondly, the larger loop by definition possesses a greater transmitter moment ($=$ turns $\times$ cross-sectional area $\times$ current).

Offsetting these advantages of using a larger loop source as the transmitter is the opposing factor that a loop couples best with a body of approximately the same size and thus only very large targets would be optimally excited. Theoretical computations by Lajoie and West (1976) confirm this analysis in a general way but indicate that the optimum source size depends also upon the overburden and host rock resistivities as well as the size of the conductivity inhomogeneity. These results are confirmed by transient electromagnetic (TEM) model studies of the Elura, New South Wales, massive sulphide deposit (Spies, 1980). Spies notes that in resistive environments a 130 m loop gave signal-to-noise ratios about three times higher than a 60 m loop over the pipe-like Elura orebody, which has a diameter of about 100 m. In conductive environments, the choice of loop size depends on the relative response expected from the host and the body and thus depends on the depth and conductivity of the body as well as the conductivity of the host rocks. Spies (1982) notes that areas of laterally anisotropic conductive overburden may offset the advantages of using smaller loops because these will be more sensitive to small spatial wavelength inhomogeneities than will larger loops.

4. DATA PROCESSING AND INTERPRETATION

As discussed above, acquisition of broadband electromagnetic data over a real earth requires collecting measurements of a number of frequencies and/or spacings (frequency domain) or at a number of time delays following transmitter shutdown (time domain). Recent developments in microelectronics coupled with quantum steps in data interpretation have done much to enhance the practicality of acquiring, processing and analysing the large volume of data generated by broadband EM surveys. The large amount of data generated by a broadband (particularly transient or sweep-frequency) electromagnetic system is no longer a concern with current high-volume data-recording techniques.

Until the mid-1970s, collection and processing of electromagnetic survey data was almost entirely analog. Zonge (1973), Lamontagne and West (1973), Jain and Morrison (1976), Snyder (1975, 1976), Nabighian (1977), Buselli and O'Neill (1977), Hohmann et al. (1977), Duncan et al. (1980) and Jeppsen (personal communication, 1981) have presented descriptions of in-field digital processors with application to broadband electromagnetic receiver systems. Use of these in-field digital micro-processors allows a wide range of pre-programmed software applications such as smoothing, filtering, stacking via coherent harmonic detection, plus spectral storage and spectral weighting (Ward, 1979). The use of magnetic recording media such as cassette recorders or floppy disks in conjunction with programmable calculators or microcomputers allows not only for in-field data verification and editing, but also for preliminary interpretation.

While advances in broadband electromagnetic instrumentation and data collection have taken major steps in the past five to ten years, the interpretational methodology has proved to be more difficult. As Ward (1979) has pointed out, realistically, any base metal environment is three-dimensional (3-D) in that any geoelectric parameter can vary significantly over relatively short distances along any of the three axes. 3-D models of the earth must include all the elements of the earth discussed above, such as massive sulphides, disseminated sulphides, conductive overburden, host rock, weathered host rock, faults or other structures, plus buried surface topography.

Details of the interpretation of broadband electromagnetic survey results are beyond the scope of this chapter. In subsequent paragraphs I hope to suggest directions that interpretation is taking and provide more detailed references to which the interested reader can refer for details.

Interpretation of electromagnetic sounding or profiling data can be accomplished using either forward or inverse methods. The forward method has received much wider application and consists of attempting to match visually observed (survey) data with predicted data obtained by mathematical or scaled physical modelling. Plots are made, ideally at the same scale, for both observed and predicted data of the real and imaginary parts of field components (or of tilt angle and ellipticity) versus induction number, frequency, or transmitter–receiver separation. This approach necessitates extensive sets of master curves for 'typical' earth models.

With the inverse model, a computer algorithm is used to compare observed field data with numerically generated values. Using an iterative process, the computer algorithm minimises the least-squares difference between the field data and the generated model data until an optimum fit is generated. The inverse method avoids 'human bias' in fitting the data to a curve and provides estimates of the resolution and reliability of the theoretical model used to fit the observed data. The method also estimates the density of information in the observed data and hence permits the design of an experiment to optimise the information density (Ward *et al.*, 1974*a*). Recently, both 2-D and 3-D inversion algorithms for active source EM have been developed (Ward, personal communication, 1982). The development of efficient inversion algorithms for application to 'real world' problems is currently a field of much research.

Three approaches have been used to generate curves for use in forward modelling. These are: (1) physical scaled modelling, (2) analytic solutions and (3) numerical solutions.

Physical scaled modelling has until recently provided the main basis for the interpretation of electromagnetic profiling data. This modelling technique makes use of the scaling identity equation:

$$(\sigma\mu\omega)^{1/2}S = (\sigma'\mu'\omega')^{1/2}S'$$

where σ represents the conductivity, μ is the magnetic permeability, ω is the frequency and S is the dimension. The primed values refer to the model while the unprimed values are for the earth inhomogeneity. Initially, sheets of metal were suspended in air. All other elements of the earth were ignored (Ward, 1971). Sarma and Maru (1971) and Verma (1972) conducted pioneering experiments utilising salt water to simulate a conductive environment. Ward *et al.* (1974*b*) reviewed scaled modelling results and suggested some of the difficulty of using scaled physical models, such as adequately representing the surface impedance at material–material interfaces. Ward (1979) has suggested that the cost and difficulty of numerically or analytically modelling complex 3-D environments may force us to rely to a greater extent on scaled physical models, perhaps using imagination and newly available materials such as graphitic resins.

Analytic solutions use fundamental equations of electromagnetism to solve theoretically for the fields due to a given conductor. Unfortunately, analytic solutions are only available for the most simplistic models and thus are highly limited in their application to 'real earth' environments. As summarised by Ward (1979), analytic solutions now exist for a number of simple models, including:

1. electric or magnetic dipoles over a homogeneous earth (Wolf, 1946; Wait, 1953, 1955; Quon, 1963; Pridmore, 1978);
2. electric or magnetic dipoles over a layered earth (Wolf, 1946; Wait, 1958; Quon, 1963; Frischknecht, 1967; Ward, 1967);
3. a uniform alternating magnetic field incident upon a sphere or cylinder in free space (Wait, 1951, 1952; Negi, 1962; Ward, 1967);
4. magnetic dipoles near a spherical body in free space (Nabighian, 1971; Lodha and West, 1976; Best and Shammas, 1978); and
5. magnetic dipoles near a sphere in a conductive half-space (Singh, 1973).

The analytic solutions are typically quite economical, in terms of computer time, and can serve to check the validity of results obtained from the more general numerical solutions.

Numerical solutions can be obtained for much more complex 2-D and 3-D models than is possible using analytic solutions. In fact, almost completely arbitrary conductivity distributions can be modelled using the numerical approach. Four methods have been used during the 1970s to compute electromagnetic scattering due to 2-D and 3-D conductivity inhomogeneities. These are: (1) finite difference, (2) finite element, (3) network analogy and (4) integral equation. In addition, a hybrid numerical equation method combining the finite element and integral equation methods has recently been reported by Lee *et al.* (1981). Based on a concept by Scheen (1978), this hybrid method is substantially more efficient than previous methods.

Applications and development of these numerical methods to 'real earth' problems previously not amenable to analysis have been the subject of many recent papers. Coggon (1971), Hohmann (1971), Parry and Ward (1971), Swift (1971), Vozoff (1971) and Dey and Morrison (1973) dealt with 2-D inhomogeneities in the vicinity of line sources. Stoyer and Greenfield (1976) describe a 2-D inhomogeneity in the field of a 3-D source. Articles by Lee (1974), Hohmann (1975), Lajoie and West (1976), Pridmore (1978) and Pridmore *et al.* (1979) describe full 3-D electromagnetic scattering problems. Hohmann (1977) contains an excellent overview of the state-of-the-art of interpretation of resistivity, induced polarisation and electromagnetics. Pridmore *et al.* (1981) present a comprehensive analysis of the application of the finite element method to modelling electromagnetic results.

The main drawback to the numerical modelling method is the cost. Ward (1979) illustrated that a 3-D modelling problem using the finite element might cost about US$300. Application of this approach to broadband methods can become prohibitively expensive. Costs are expected to fall significantly through the development of more efficient algorithms, and the use of dedicated minicomputers, particularly in conjunction with linear array processors, or even the development of a computer designed solely for solving numerical EM scattering problems.

5. CASE HISTORIES

As an illustration of the application of broadband electromagnetic systems to 'real earth' problems, the results of two recent surveys are presented. The results are largely frequency domain; this reflects no bias as to the superiority of this domain, but only the availability of more frequency-domain results on specific properties with which the author is acquainted.

Additional case histories of broadband electromagnetic surveys (as well as other geophysical methods) have recently been published for two Australian deposits, the Elura deposit (Emerson, 1980) and the Woodlawn deposit (Whiteley, 1981), both in New South Wales. These two volumes present both time-domain and frequency-domain survey results for two very different geological environments and are worth close review by the applied explorationist.

5.1. 'Western World' Massive Sulphide Deposit, Yuba–Nevada Counties, California

Pridmore *et al.* (1979) reported on broadband electromagnetic measurements over a massive sulphide prospect in the California Foothills copper–zinc belt. Additional detail on the EM survey is given by Pridmore (1978). The deposit contains approximately 2.5 million tons of copper-bearing pyritic ore and is located near the Yuba–Nevada County border about 3 miles south of the hamlet of Smartsville, California.

The deposit is located within the Smartsville ophiolite; the massive sulphides and a halo of disseminated, barren sulphides occur as a stratiform horizon within the ophiolite complex. The massive sulphides are intimately associated with a variably altered intermediate (andesite–

dacite) tuff unit, locally designated as 'pyritic felsic tuff'. A schematic geological cross-section of the deposit is shown in Fig. 12. The surface geology in the vicinity of the deposit is shown in Fig. 13 (from Pridmore *et al.*, 1979). The region has been subjected to lower greenschist facies metamorphism. The deposit occurs on and near the axial plane of the eastern flank of a major regional structure, the Timbuctoo anticline; thus, dips in the area of the deposit are 10–20° to the north-east. The strike of the geology in the area is about 15°N–20°W. Additionally, the massive sulphide lenses have about a 10° plunge to the north. A north–south trending fault runs through the area and is reflected geomorphologically by the presence of a swampy stream channel over portions of the deposit; this swamp is developed best to the north of line 18S and varies from 9 to 15 m in thickness.

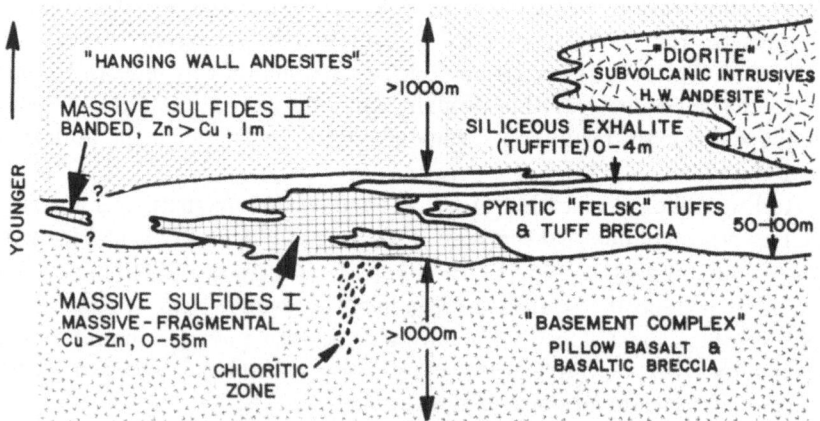

FIG. 12. Schematic geological cross-section of the Western World massive sulphide deposit, California Foothills, showing the complex shape of the massive sulphides and a 'halo' of pyritic tuffs.

The massive sulphides are contained primarily in two irregular lenses which grade easterly into a variably disseminated zone within the tuffaceous horizon. A near-vertical, east–west fault through line 18S downdrops mineralisation on the north side of the fault to depths of 40 m or more. The massive sulphides vary in thickness from 1 m to over 50 m and are generally concordant with bedding. Figure 13 shows the projection of the 20-ft massive sulphide contour to the surface of the earth as a heavy dotted line. Approximately 120 diamond drill holes from 100 to

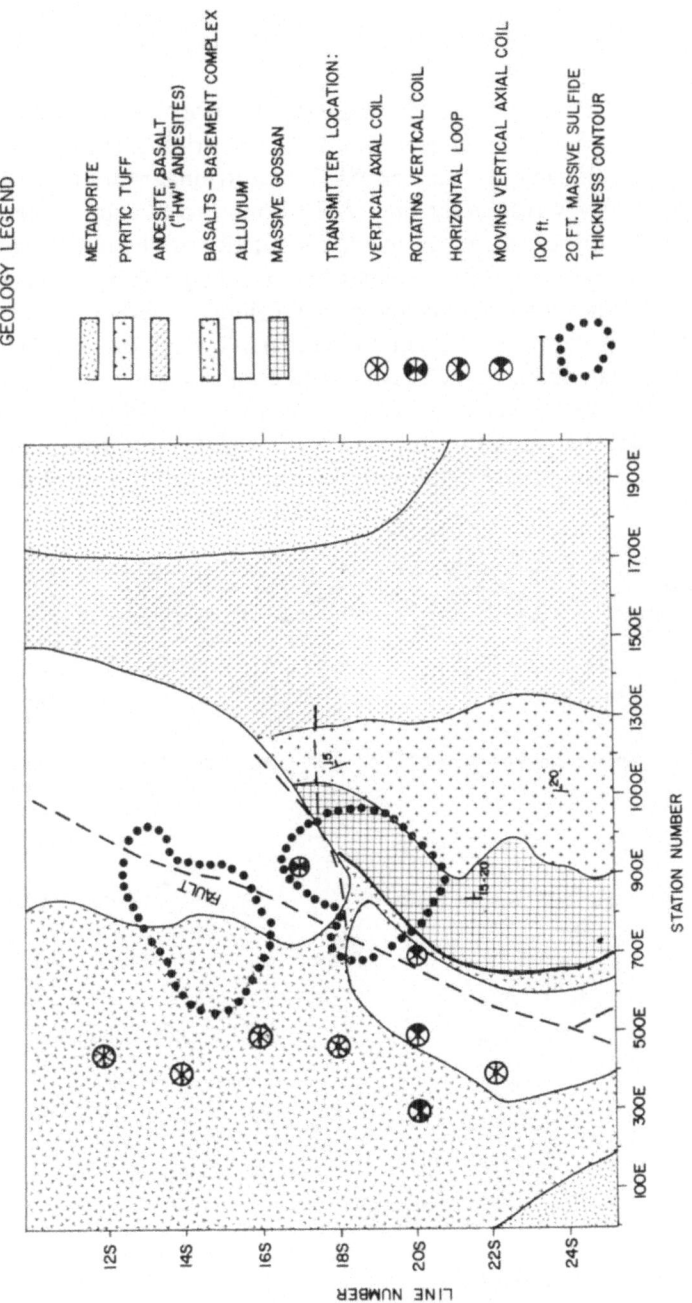

Fig. 13. Geological map of the Western World massive sulphide deposit, California Foothills. The vertical projection of the 20-ft thickness contour to the surface is shown with heavy dotted lines.

500 ft in depth have been drilled to date on the property and provide excellent geological control.

Discovery in 1973 of pyritic tuff and gossan at the property resulted in initial geophysical and geochemical surveys. A pre-drilling Turam survey was conducted utilising a Scintrex SE-71 electromagnetic unit. The 400 Hz Turam survey results for lines 22S through to 14S are shown in Fig. 14. A moderate response over massive sulphides can be noted on the line 20S data; weaker responses on lines 18S, 16S and 14S were believed to be caused by the swamp. After the initial drilling (hole No. 4 was the discovery hole on the project) encountered massive sulphides, but before their geometry was well known, a broadband electromagnetic survey was conducted using the University of Utah's 14-frequency EM unit. Results of this survey are discussed by Pridmore *et al.* (1979). Three transmitter–receiver configurations were employed: vertical axial loop, vertical rotat-

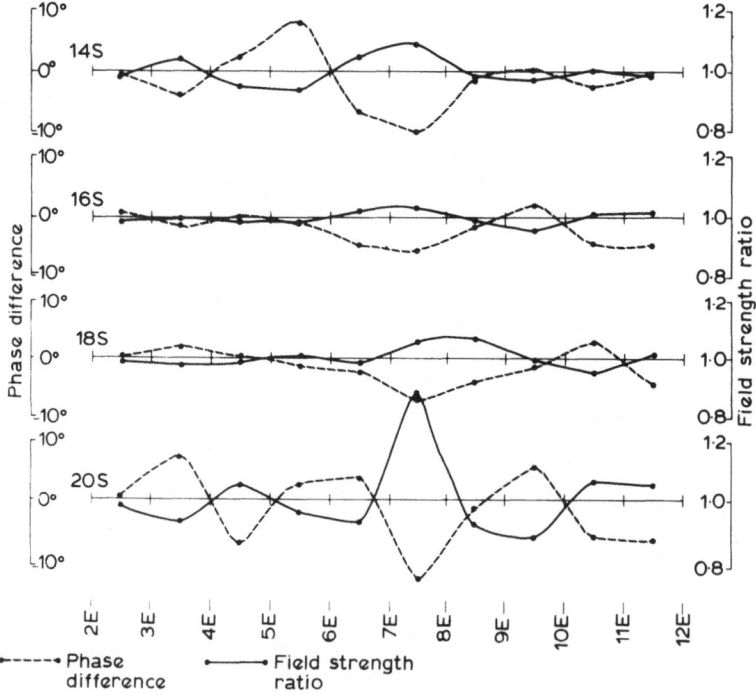

FIG. 14. 400 Hz Turam results obtained over the Western World massive sulphide deposit. The receiver coils utilised a 100 ft separation. The transmitter was a 4000 × 3000 ft loop with the baseline at station 0 East.

ing loop and horizontal loop. As noted therein, the response of a complex earth was resolved and responses attributable to massive sulphides, disseminated sulphides and host rock could be discriminated when tilt angle and ellipticity values were contoured in frequency–distance space.

Figure 15 shows theoretical tilt angle and ellipticity responses expected over a homogenous half-space of resistivity $100\,\Omega\,m$, contoured in frequency–distance space. Figure 16 shows the expected anomalous responses of a simple vertical dike conductivity inhomogeneity. In a real earth environment such as at the Western World deposit, where massive sulphides are present together with disseminated sulphides, a weathered zone and conductive host rocks, tilt angle and ellipticity responses combining the diagnostic responses of Figs 15 and 16 are to be expected.

Figure 17 shows the contours of tilt angle plotted in frequency–distance space for line 20S at the Western World deposit. (All the tilt angle results were normalised to the 10.5 Hz reading to remove first-order topographic effects. Without this normalisation, changes in tilt angle caused by elevation differences would obscure the subtle topographic effects produced by the massive sulphides.) Figure 17 illustrates the half-space response (trend A–A') superimposed on the response of an inhomogeneity of relatively high conductivity–thickness product (trend B–B') plus the response of an inhomogeneity of relatively low conductivity–thickness product (trend C–C') for a vertical loop source. Trend B–B' is caused by the massive sulphides of the southern lens, while trend C–C' is caused by the disseminated non-economic sulphides which grade laterally into the massive sulphides. Figure 18 shows the contours of ellipticity over the same line using the vertical loop source. It is significant to note that even though the response due to the massive sulphides is of small amplitude owing to relatively poor flux coupling, about 1–2° of tilt angle and 1–2% ellipticity, the response from the massive sulphide is clearly definable (Pridmore et al., 1979). This response magnitude is about 5–10 times the system noise level of broadband EM parameters (Ward et al., 1974b). As is typical of inductive response, the ellipticity or out-of-phase response begins at a lower frequency than the tilt angle response; in fact, the ellipticity response is recognisable all the way down to 10.5 Hz, the system's lowest. The asymmetry in the ellipticity contours is indicative of a shallow easterly dip. This shallow dip was predicted by Ward (personal communication, 1974), when initial geological drilling and mapping suggested a 45–60° easterly dip; subsequent drilling confirmed the shallow dip. At the lower frequencies, only the gross structure

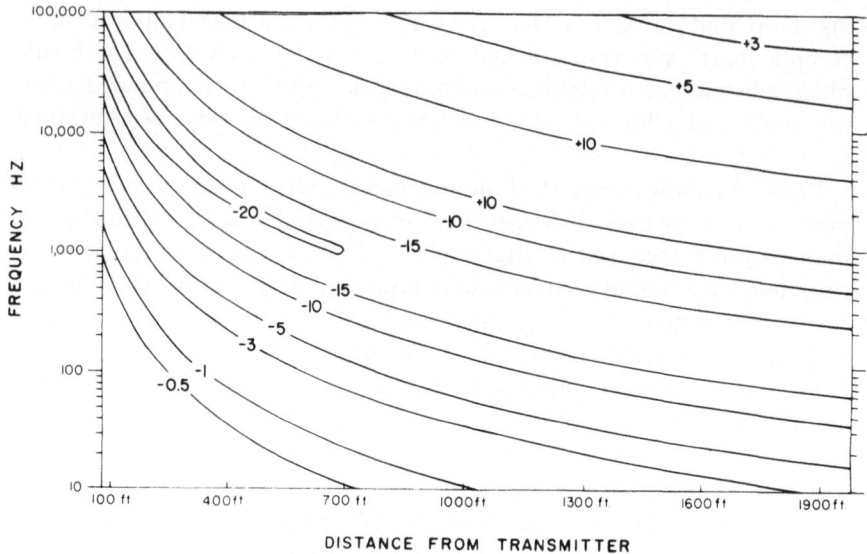

DISTANCE FROM TRANSMITTER

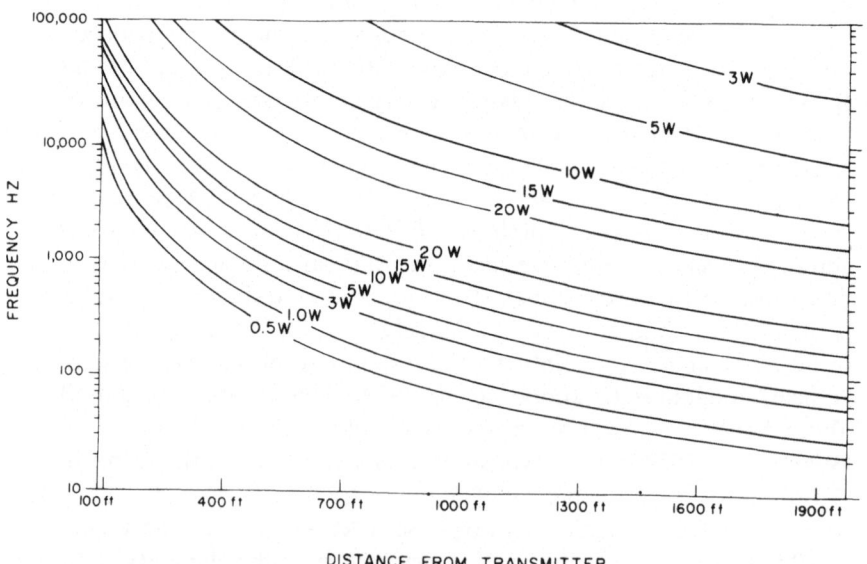

DISTANCE FROM TRANSMITTER

FIG. 15. Contours in frequency–distance space for a homogeneous half-space of resistivity 100 Ωm produced by a vertical loop transmitter: (above) ellipticity in %; (below) tilt angle in degrees. (W denotes a tilt to the west, or towards the transmitter located at 0.)

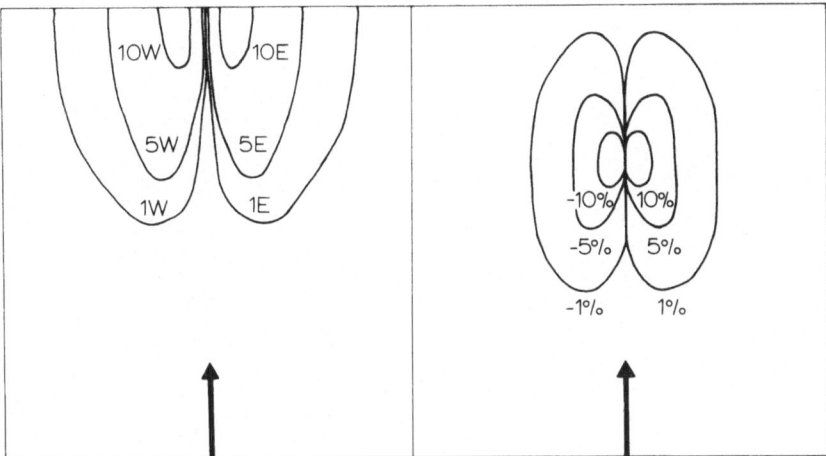

FIG. 16. Schematic depiction of contours of tilt angle (left) and percent ellipticity (right) produced by a simple dike in free space. The vertical axis is frequency, the horizontal axis is distance from the transmitter. The arrow shows the location of the inhomogeneity.

of the massive sulphide conductor is resolved. At frequencies above 1000 Hz for tilt angle and 600 Hz for ellipticity, the response due to the massive sulphides is obliterated by the host rock response.

Line 18S crosses the east–west trending fault at a very low angle, and thus complicates the presentation of the geological section shown in Fig. 19. On this line, the ellipticity response of the massive sulphides is centred around station 9 + 50. It is smaller than on line 20S and is not characterised by marked asymmetry. Also, the tuff response on this line is considerably less than the response on line 20S. Overall, the half-space response is much stronger than on the line 20S ellipticity data.

The effect of transmitter–receiver separation on anomaly amplitude over the Western World deposit was examined by Pridmore et al. (1979) and Pridmore (1978). Figure 20 shows the results of moving the vertical loop transmitter from station 3E to 7E on line 20S. The axes are percent ellipticity and degrees of tilt angle, which for small responses are nearly percent quadrature and percent in-phase of the secondary vertical field with respect to the primary horizontal field (Grant and West, 1965, p. 484). The results show that peak anomaly amplitude increases and the ellipticity changes sign at successively higher frequencies as the distance between the transmitter and the massive sulphides decreases. The in-

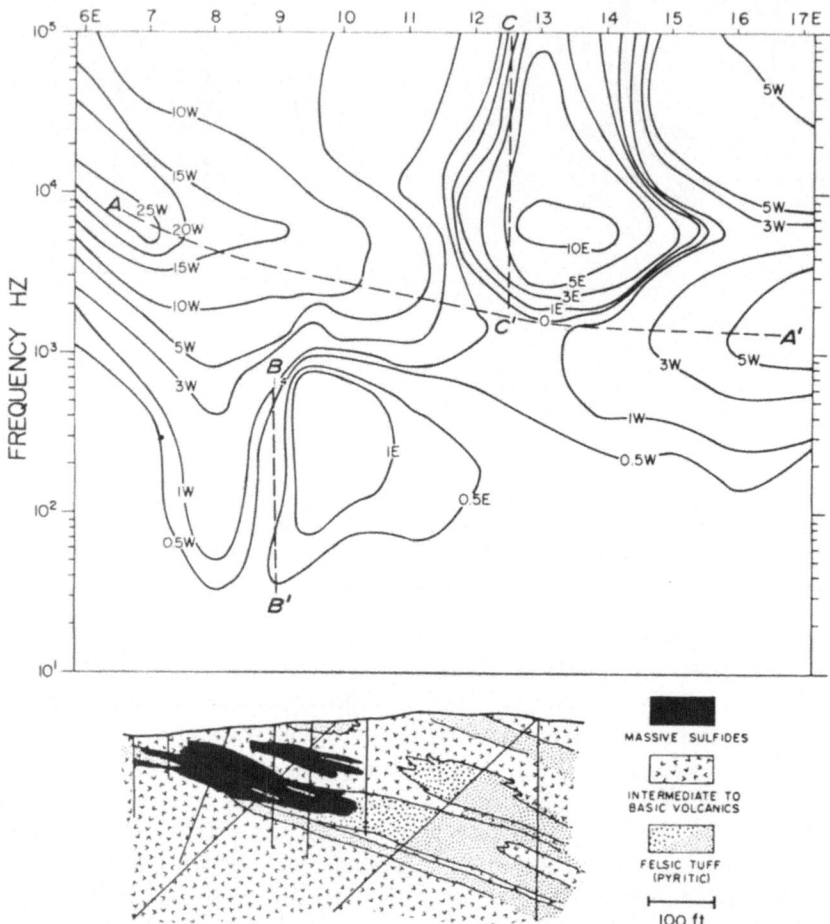

FIG. 17. Contours of tilt angle in frequency–distance space at the Western World, California, deposit line 20S. Trends A–A', B–B' and C–C' represent the response of half-space, massive sulphides and disseminated sulphides, respectively. Vertical axial loop source normalised to 10.5 Hz, Tx at 3 + 00E.

crease in anomaly amplitude is at least partially due to improved flux coupling (see Pridmore *et al.*, 1979, Fig. 13a).

At the Western World deposit, Pridmore *et al.* (1979) concluded that (1) the results from the vertical axial loop transmitter are easier to interpret, in terms of simple models, than those produced by the horizontal loop transmitter; and (2) the vertical rotating loop produced a larger

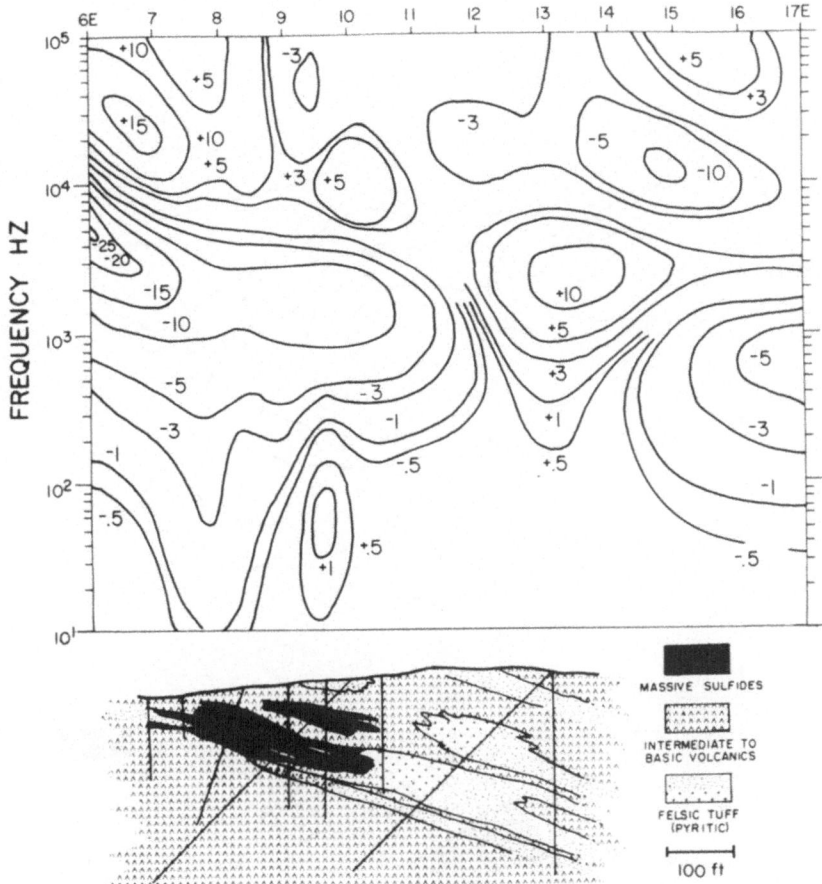

Fig. 18. Contours of percent ellipticity in frequency–distance space at the Western World, California, deposit line 20S. Vertical axial loop source, Tx at 3+00E.

amplitude response from the massive sulphides, but had no more resolution in terms of fine structure in the massive sulphides than did the vertical axial loop transmitter.

Figures 21 and 22 show the results of a commercially available horizontal loop EM unit, the Apex Parametrics Max-Min II EM, over lines 20S and 18S respectively. The largest response occurs on line 20S, where the massive sulphides are both thicker and closer to the surface than on line 18S. The two conductors present near 600E (the same

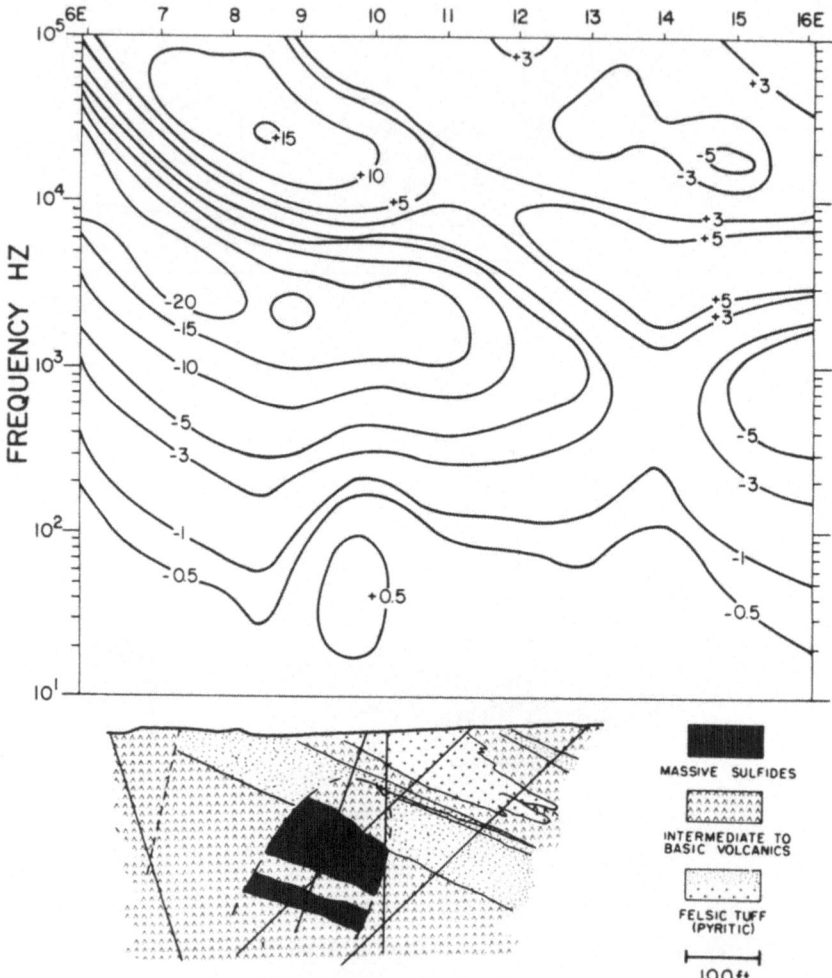

FIG. 19. Contours of percent ellipticity in frequency-distance space at the Western World, California, deposit line 18S. Vertical axial loop source, Tx at 3+00E.

location as 6E) and 1000E apparently represent the edges of the re-latively flat-lying massive sulphide lens. Curve matching suggests that the body is almost horizontal; it appears that current gathering due to the conductive host rocks may be playing a role in minimising the asymmetry of the response. Less separation of the responses due to the

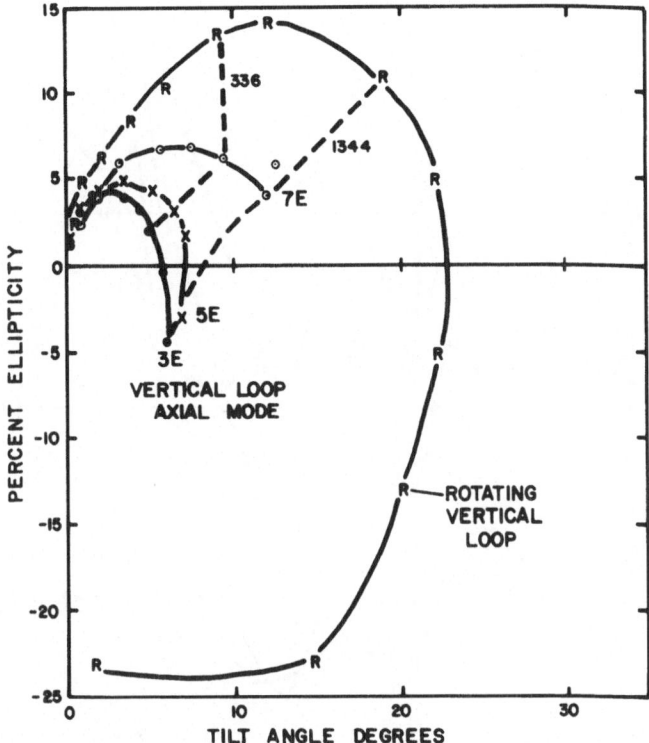

FIG. 20. Anomaly amplitudes over the massive sulphides on line 20S plotted in pseudo-phasor form for both vertical loop configurations, viz. vertical loop transmitter sites on line 20S at eastings of $3+00E$ (●), $5+00E$ (×), $7+00E$ (○) and the rotating vertical loop transmitter on line 17S at $9+24E$ (R). The anomaly could be identified at frequencies of 10.5, 21, 42, 84, 168, 336, 672 and 1344 Hz for data from all configurations. In addition, for the rotating vertical loop transmitter the anomaly could be identified at frequencies of 2688, 5376, 10 752, and 43 008 Hz. Results at frequencies of 336 and 1344 Hz are joined with dashed lines.
(From Pridmore *et al.*, 1979.)

massive sulphides and the disseminated sulphide is produced with the horizontal loop EM unit than with the 14-frequency system with the vertical axial loop mode.

5.2. Cavendish Test Site, Ontario, Canada

The Cavendish test site is located in Cavendish Township, Ontario, approximately 6 miles south of the village of Gooderham and about 100 miles north-east of Toronto. The site was selected by the Geological

J. W. MOTTER

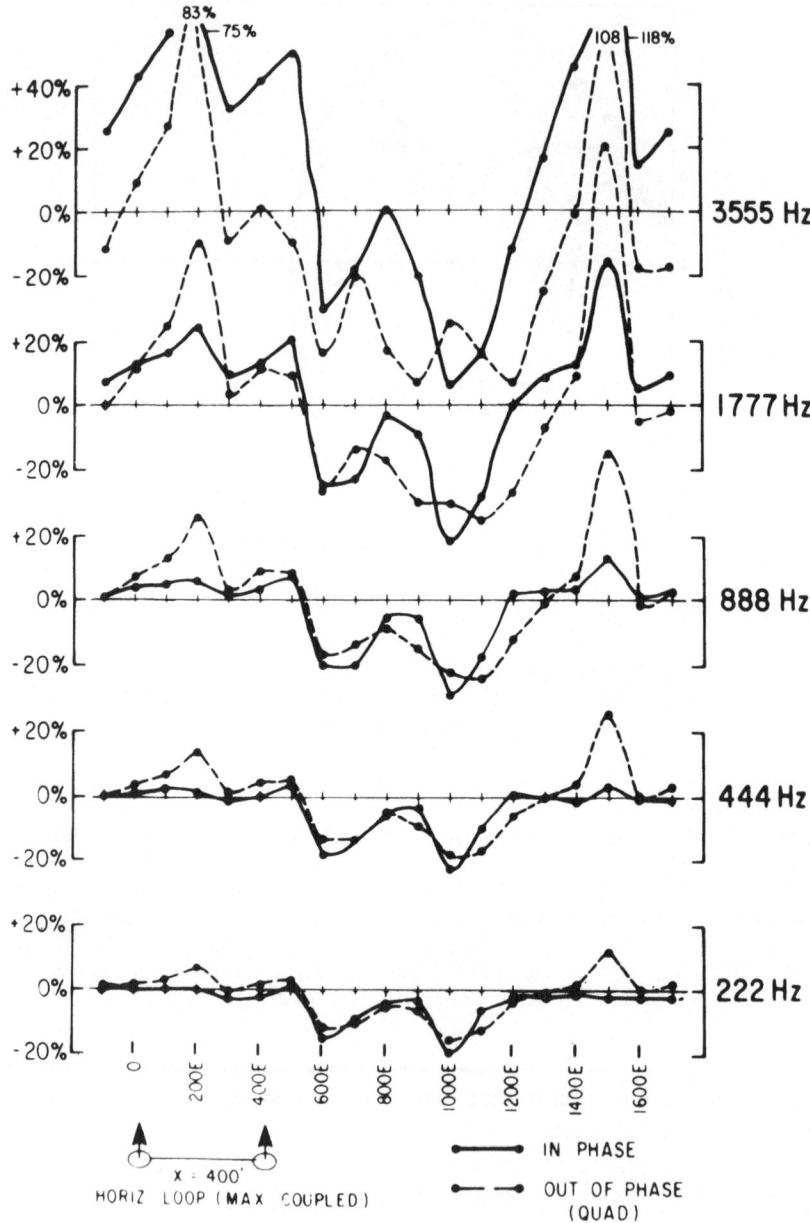

FIG. 21. Apex Max–Min II EM survey results over line 20S at the Western World, California, deposit.

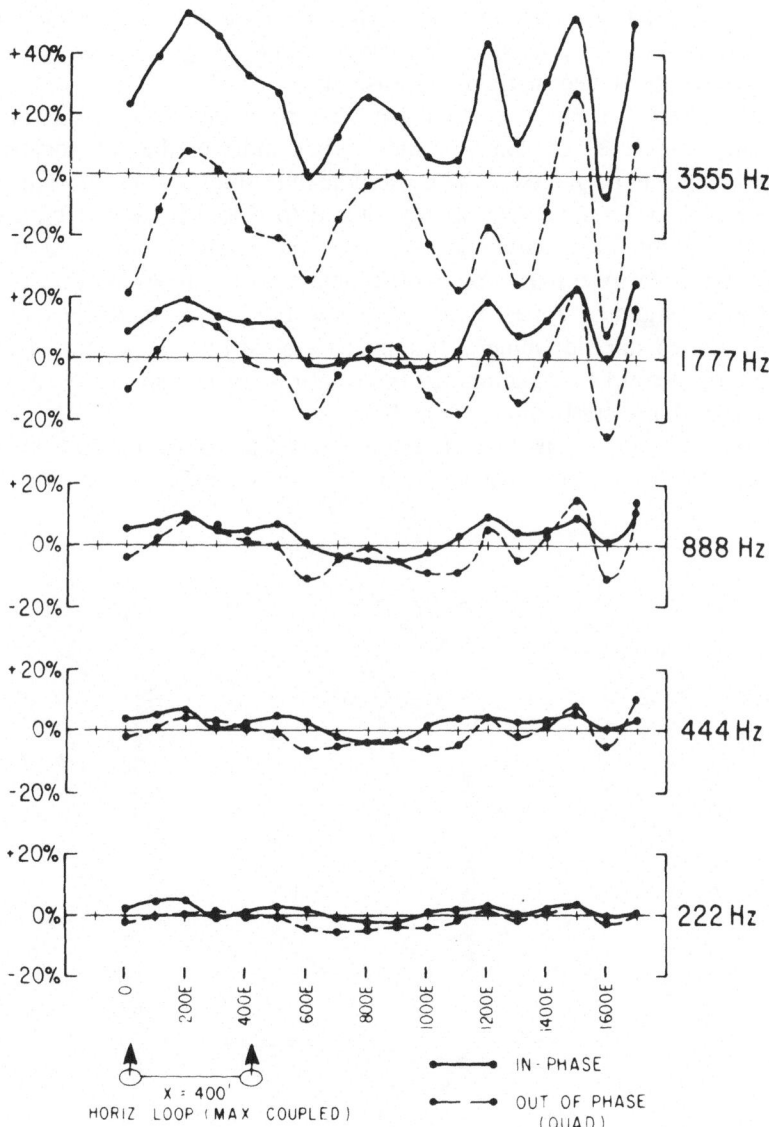

FIG. 22. Apex Max-Min II EM survey results over line 18S at the Western World, California, deposit.

Survey of Canada in 1967 as a test location for evaluation of geophysical techniques and equipment; consequently, numerous geophysical surveys have been conducted at the Cavendish site.

The geology of the Cavendish site has been described in detail by Williams *et al.* (1973). The two major rock units of the site consist of granitic and mafic gneisses, and crystalline limestone of Precambrian age. These rock units have been highly altered to calc–silicate and typically contain about 0.1% disseminated pyrite and pyrrhotite mineralisation. Two zones of heavier pyrite–pyrrhotite mineralisation are present at the site and are designated zones A and B. These two zones exhibit a core of concentrated sulphide mineralisation surrounded by a disseminated halo containing from 1 to 2% sulphides. Information suggests an easterly dip for zone A and a westerly dip for zone B.

Figure 23 shows a geological section along Cavendish line C. Williams *et al.* (1973) report that drilling on line C showed zone A to consist of a

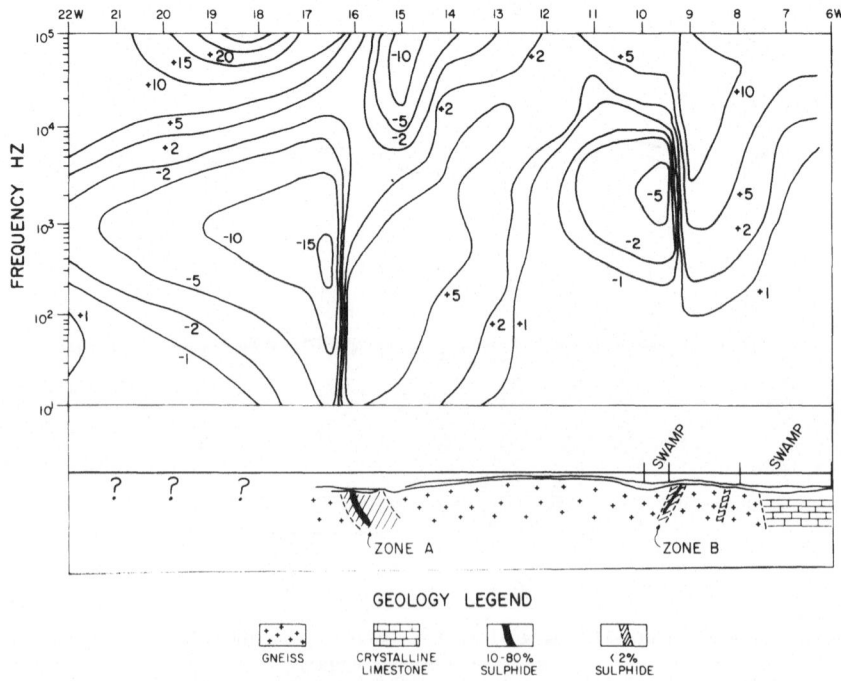

FIG. 23. Contours of percent ellipticity in frequency–distance space for Cavendish test site line C. Vertical axial loop source, Tx at 4+00W. (From Ward *et al.*, 1974*b*.)

core zone containing 10 ft of pyrite–pyrrhotite stringers enclosed in an 80-ft halo of 1–2% sulphides. Zone B near station 9+00W consists of 68 ft of 2% sulphides which contain two small veins, 4 inches and $\frac{1}{4}$ inch thick, of 80% sulphides.

Also shown in Figure 23 are contours of percent ellipticity in frequency–distance space (from Ward *et al.*, 1974*b*). The results presented are for the vertical axial loop configuration discussed previously. Zones A and B are clearly delineated in this figure; it is also apparent that zone A has a much higher conductivity–thickness product because the responses start at lower frequencies. Unlike the ellipticity contours in frequency–distance space for the Western World deposit (see, e.g., Figs 18 and 19), the effects of overburden, host rock and disseminated halo are not as apparent. Figure 24, however, clearly reveals that either the host rock or

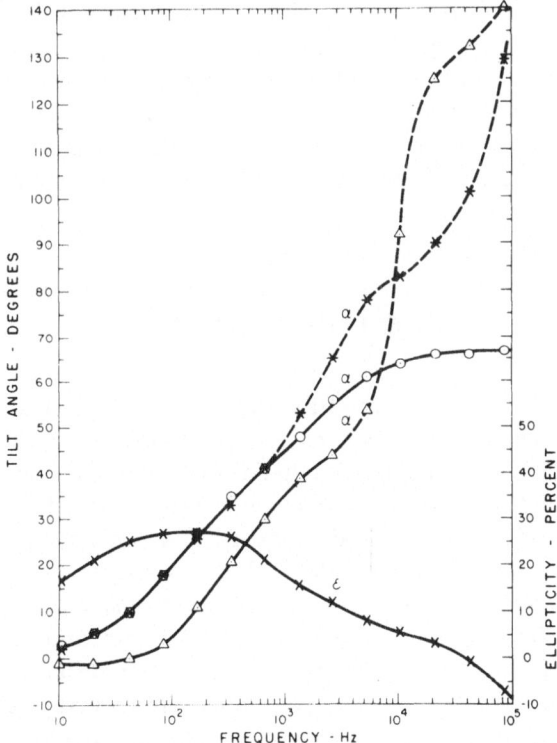

FIG. 24. Peak-to-peak tilt angle, α ($\odot$) and ellipticity, ε ($\times$) for the vertical axial coil; peak-to-peak tilt angle (∗) for the vertical rotating coil; and peak tilt angle ($\triangle$) for the horizontal coil. All are plotted against frequency for zone A of the Cavendish test site. (After Ward *et al.*, 1974*b*.)

the overburden is contributing to the response at frequencies above
10 kHz at the Cavendish A zone. As shown in this figure, with the
vertical axial coil source the tilt angle exhibits both asymptotes over four
decades of frequency, while the ellipticity at 100 kHz is at the high-
frequency asymptote but fails to reach a low-frequency asymptote. The
horizontal loop and vertical rotating loop sounding curves are not at all
smoothly varying, indicative of complex interaction between two or more
elements of the geoelectric section.

Figures 25 and 26 present profiles of tilt angle and ellipticity across
Cavendish line C. These profiles, from Ward et al (1974b), also demonstrate the higher conductivity–thickness product of zone A. Unlike
the contours of tilt angle in frequency–distance space, these profiles have
not been normalised to the 10.5 Hz tilt angle data to remove first-order
topographic effects. The inferred dips of the A and B zones from the tilt
angle profiles are consistent with dips obtained from drilling results.

Figure 27 presents early time-domain results obtained by Lamontagne
and West (1973) over line C at the Cavendish test site. A large horizontal
loop was used as the transmitter. These time-domain results demonstrate
that zone A has a higher conductivity–thickness product than zone B.
These results are in agreement with the frequency-domain results presented earlier.

6. CONCLUSIONS

In this chapter we have examined some of the geological complexities of
a 'real earth' environment. We have noted that the components of the
earth in the vicinity of a massive sulphide deposit will include not just the
massive sulphides but may also include a disseminated halo, conductive
host rock, conductive overburden, buried and surface topography, and
faults or shear zones. A broadband EM survey attempts to excite each of
these components. Current broadband electromagnetic systems, and
those yet to be developed, will allow us to obtain information on the
parameters of a real three-dimensional earth. The variables which must
be determined are numerous and complexly interrelated. Although much
has been accomplished in terms of interpretation and modelling of
complex inductive responses in the past decade, much more still remains.
It appears that promising avenues for significant gains in modelling

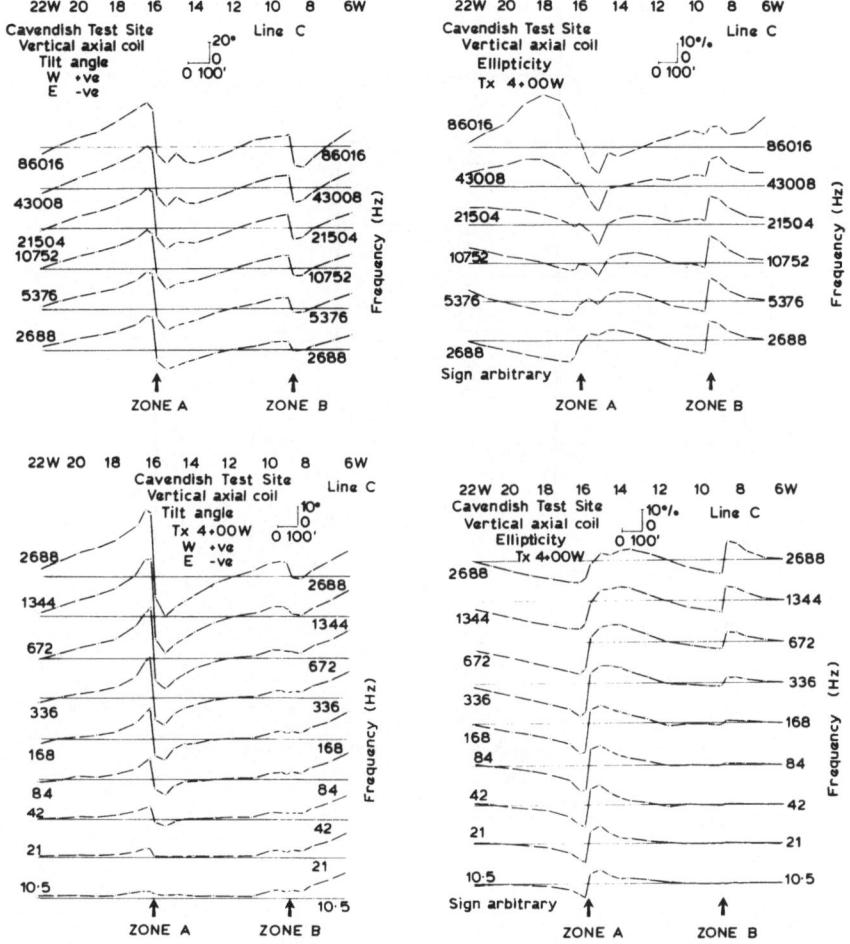

FIG. 25. Tilt angle profiles from Cavendish line C, vertical axial loop source. (From Ward *et al.*, 1974*b*.)

FIG. 26. Ellipticity profiles from Cavendish line C, vertical axial loop source. (From Ward *et al.*, 1974*b*.)

complex electromagnetic responses may be derived from the application of dedicated minicomputers such as the VAX and the Prime to solving forward modelling problems. Some installations of this sort are already being used (Ward, personal communication, 1981). The use of 'space age' materials and technology to create new types of scaled physical models may also be productive.

The combination of instrumental advances coupled with full 3-D

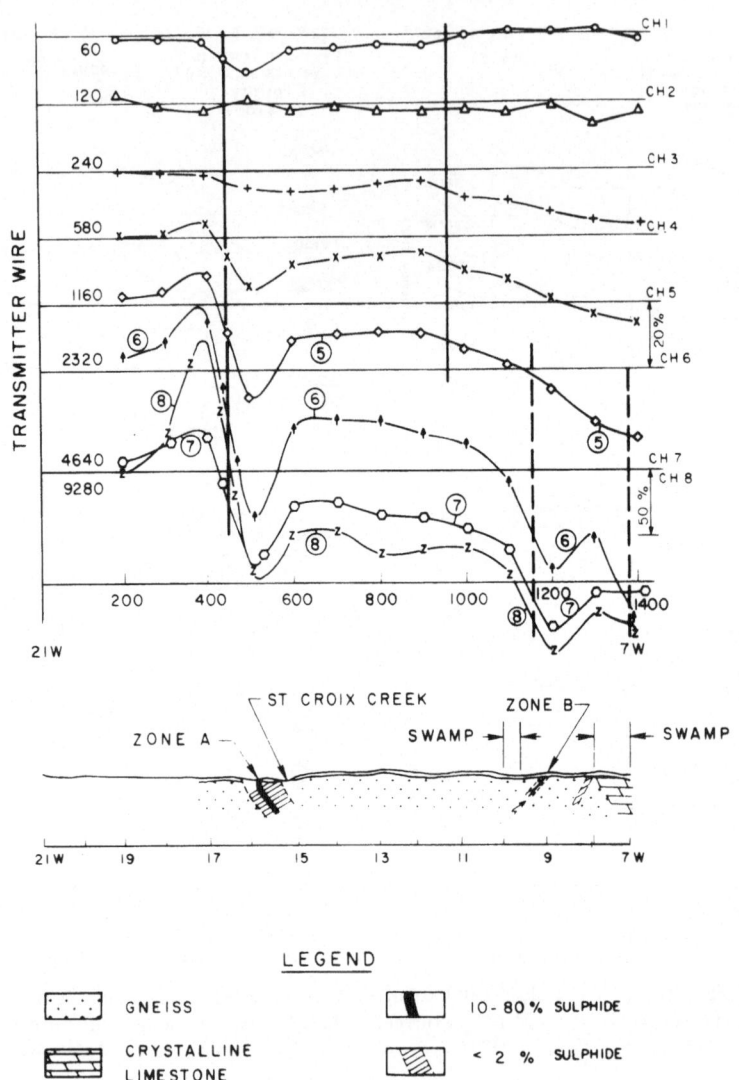

FIG. 27. Profiles of 7-channel UTEM time-domain response on line C of Cavendish test site. (After Lamontagne and West, 1973.)

interpretational capability on digital computers will significantly enhance the usefulness of the broadband electromagnetic method in complex geological environments. This is necessary if ore deposits are to be located at increasing depths in locations throughout the world.

ACKNOWLEDGEMENTS

Numerous individuals and firms have provided help in supplying data for this chapter. Their assistance is gratefully acknowledged. A. A. Fites made valuable suggestions as to the structure of the chapter. The author wishes to extend special thanks to Stan Ward, who encouraged the writing of this chapter, critically reviewed an early draft of the manuscript, and who provided original artwork for many of the figures, as well as many helpful discussions and suggestions. Whitney & Whitney, Inc., made time available for writing; John Whitney, Ron Whitney and Vicki Erickson reviewed the manuscript and made valuable suggestions. Kris Covington and Rosalie Carpenter carefully drafted the artwork. Anne Conibear patiently typed numerous revisions of the manuscript.

REFERENCES

BEST, M. E. and SHAMMAS, B. R. (1978) A general solution for a spherical conductor in a magnetic dipole field. Preprint, Shell Canada Resources Ltd.

BRAHAM, B., HAREN, R., LAPPI, D., LEMAIRE, H., PAYNE, D., RAICHE, A., SPIES, B. and VOZOFF, K. (1978) Lecture notes from the US–Australia electromagnetic workshop. *Bull. Aust. Soc. Explor. Geophys.* **9**(1), 2–33.

BUSELLI, G. and O'NEILL, B., (1977) Sirotem: a new portable instrument for multichannel transient electromagnetic measurements. *Bull. Aust. Soc. Explor. Geophys.* **8**(3), 1–6.

COGGON, J. H. (1971) Electromagnetic and electrical modelling by the finite element method. *Geophysics* **36**(1), 132–55.

CRONE, J. D. (1966) The development of a new ground EM method for use as a reconnaissance tool. In *Mining Geophysics*, Vol. 1, Society of Exploration Geophysicists.

CRONE, J. D. (1973) Model studies with the Shootback method. In *Proceedings of the Symposium on Electromagnetic Exploration Methods*, University of Toronto.

DEY, A. and MORRISON, H. F. (1973) Electromagnetic response of two-dimensional inhomogeneities in a dissipative half-space for Turam interpretation. *Geophys. Prosp.* **21**, 340–65.

DEY, A. and MORRISON, H. F. (1977) Resistivity modelling for arbitrarily shaped three-dimensional structures. Lawrence Berkeley Lab., Preprint LBL-7010.

DEY, A. and WARD, S. H. (1970) Inductive sounding of a layered earth with a horizontal magnetic dipole. *Geophysics* **35**(4), 660–703.

DUNCAN, P. M., HWANG, A., EDWARDS, R. N., BAILEY, R. C. and GARLAND, G. D. (1980) The development and applications of a wide band electromagnetic sounding system using a pseudo-noise source. *Geophysics* **45**(8), 1276–96.

EMERSON, D. W. (ed.) (1980) *The Geophysics of the Elura Orebody: Proceedings of the Elura Symposium*, Australian Society of Exploration Geophysicists, Sydney.

FRISCHKNECHT, F. C. (1967) Fields about an oscillating magnetic dipole over a two-layer earth, and application to ground and airborne electromagnetic surveys. *Q. Colo. School Mines* **62**(1), 326pp.

GAUR, V. K. and VERMA, O.P. (1973) Enhancement of electromagnetic anomalies by a conducting overburden II. *Geophys. Prosp.* **21**(1), 159–84.

GAUR, V. K., VERMA, O. P. and GUPTA, C. P. (1972) Enhancement of electromagnetic anomalies by a conducting overburden. *Geophys. Prosp.* **20**(3), 580–604.

GLENN, W. E. and WARD, S. H. (1976) Statistical evaluation of electrical sounding methods. Part I: Experiment design. *Geophysics* **41**(6A), 1207–21.

GRANT, F. S. and WEST, G. F. (1965) *Interpretation Theory in Applied Geophysics*, McGraw/Hill, New York.

HOHMANN, G. W. (1971) Electromagnetic scattering by conductors in the earth near a line source of current. *Geophysics* **36**(1), 101–31.

HOHMANN, G. W. (1975) Three-dimensional induced polarization and electromagnetic modelling. *Geophysics* **40**(2), 309–24.

HOHMANN, G. W. (1977) Modelling team report, Workshop on electrical methods in geothermal exploration. US Geol. Surv. Grant 14-08-0001-G-359, University of Utah, January 1977.

HOHMANN, G. W., NELSON, P. H. and VAN VOORHIS, G. D. (1977) A vector EM system and its field applications. *Geophysics* **43**(7), 1418–40.

HOOD, P. (1977) Mineral exploration trends and developments in 1976. *Can. Mining J.* **98**(1), 8–47.

JAIN, B. and MORRISON, H. F. (1976) Inductive resistivity survey in Grass Valley, Nevada. Lawrence Berkeley Lab., Progress Report.

LAJOIE, J. J. (1977) Two selected field examples of EM anomalies in a conductive environment. *Geophysics* **42**(3), 655–60.

LAJOIE, J. J. and WEST, G. F. (1976) The electromagnetic response of a conductive inhomogeneity in a layered earth. *Geophysics* **41**(6A), 1133–56.

LAMONTAGNE, Y. (1975) Applications of wide-band, time-domain EM measurements in mineral exploration. Unpublished Ph.D. thesis, University of Toronto.

LAMONTAGNE, Y. and WEST, G. F. (1973) A wide-band, time-domain ground EM system. In *Proc. Symposium on Electromagnetic Exploration Methods*, University of Toronto.

LEE, K. H., PRIDMORE, D. F. and MORRISON, H. F. (1981) A hybrid three-dimensional electromagnetic modelling scheme. *Geophysics* **46**(5), 796–805.

LEE, T. (1974) Transient electromagnetic response of a sphere in a layered medium. *Geophys. Prosp.* **23**, 492–512.

LODHA, G. S. (1977) Time domain and multifrequency electromagnetic responses in mineral properties. Unpublished Ph.D. thesis, University of Toronto, 183 pp.

LODHA, G. S. and WEST, G. F. (1976) The electromagnetic response of a conductive inhomogeneity in a layered earth. *Geophysics* **41**(6A), 1133–56.

LOWRIE, W. and WEST, G. F. (1965) The effect of a conducting overburden on electromagnetic prospecting measurements. *Geophysics* **30**(4), 624–32.

MCNEILL, J. D. (1980) Applications of transient electromagnetic techniques. Technical Note TN-7, Geonics, Ltd.

NABIGHIAN, M. N. (1971) Quasi-static transient response of a conducting permeable two-layer sphere in a dipolar field. *Geophysics* **36**(1), 25–37.

NABIGHIAN, M. N. (1977) The Newmont EMP methods. In Geophysics applied to detection and delineation of non-energy, non-renewable resources. Report on Grant AER76–80802, National Science Foundation; Dept of Geology and Geophysics, University of Utah.

NEGI, J. G. (1962) Inhomogeneous cylindrical ore body in presence of a time varying magnetic field. *Geophysics* **27**(3), 386–92.

PALACKY, G. J. (1975) Interpretation of input AEM measurements in areas of conductive overburden. *Geophysics* **40**(3), 490–502.

PARRY, J. R. and WARD, S. H. (1971) Electromagnetic scattering from cylinders of arbitrary cross-section in a conductive half-space. *Geophysics* **36**(1), 67–100.

PRIDMORE, D. F. (1978) Electromagnetic scattering of three-dimensional fields by three-dimensional earths. Unpublished Ph.D. thesis, University of Utah.

PRIDMORE, D. F., WARD, S. H. and MOTTER, J. W. (1979) Broadband electromagnetic measurements over a massive sulfide prospect. *Geophysics* **44**(10), 1677–99.

PRIDMORE, D. F., HOHMANN, G. W., WARD, S. H. and SILL, W. R. (1981) An investigation of finite-element modeling for electrical and electromagnetic data in three dimensions. *Geophysics* **46**(7), 1009–24.

QUINCY, E. H., DAVENPORT, W. H. and MOORE, D. F. (1976) Three-dimensional response maps for a new side-band induction system. *IEEE Trans. Geosci. Electr.* **GE-14**(4), 261–9.

QUON, C. (1963) Electromagnetic fields of elevated dipoles on a two-layer earth. Unpublished M.Sc. thesis, University of Alberta.

RAICHE, A. P. and SPIES, B. R. (1981) Coincident loop TEM master curves for interpretation of two-layer earths. *Geophysics* **46**(1), 53–64.

ROY, A. (1970) On the effect of overburden on EM anomalies: a review. *Geophysics* **35**(4), 646–59.

RYU, J., MORRISON, H. F. and WARD, S. H. (1970) Electromagnetic fields about a loop source of current. *Geophysics* **35**(5), 862–96.

SARMA, D. G. and MARU, W. M. (1971) A study of some effects of a conducting host rock with a new modelling apparatus. *Geophysics* **36**(1), 166–83.

SCHEEN, W. L. (1978) EMMMMA, a computer program for three-dimensional modeling of airborne electromagnetic surveys. In Proceedings of Workshop on Modeling of Electrical and Electromagnetic Methods, Lawrence Berkeley Lab., LBL-7053, p. 53.

SINGH, S. K. (1973) Electromagnetic transient response of a conducting sphere embedded in a conductive medium. *Geophysics* **38**(5), 864–93.

SNYDER, D. D. (1975) A programmable digital electrical receiver. Presented at 45th Annual Meeting, Society of Exploration Geophysicists, Denver.

SNYDER, D. D. (1976) Field tests of a microprocessor-controlled electrical receiver. Presented at 46th Annual Meeting, Society of Exploration Geophysicists, Houston.

SPIES, B. R. (1976) The transient electromagnetic method in Australia. *BMR J. Aust. Geol. Geophys.* **1**, 23–32.

SPIES, B. R. (1980) TEM model studies of the Elura deposit, Cobar, New South Wales. *BMR J. Aust. Geol. Geophys.* **5**, 77–85.

SPIES, B. R. (1982) One-loop and two-loop surveys, Elura deposit, Cobar, N.S.W. *Bull. Aust. Soc. Explor. Geophys.* (in press).

STOYER, C. H. and GREENFIELD, R. J. (1976) Numerical solutions of the response of a two-dimensional earth to an oscillating magnetic dipole source. *Geophysics* **41**(3), 519–20.

SWIFT, JR, C. M. (1971) Theoretical magnetotelluric and Turam response from two-dimensional inhomogeneities. *Geophysics* **36**(1), 38–52.

VANYAN, L. L. (1967) *Electromagnetic Depth Sounding*, selected and translated by G. V. Keller, Consultants Bureau, New York, 312 pp.

VERMA, O. P. (1972) Electromagnetic model experiments, simulating conditions encountered in geophysical prospecting. Ph.D. dissertation, University of Roorkee.

VERMA, O. P. and GAUR, V. K. (1975) Transformation of electromagnetic anomalies brought about by a conducting host rock. *Geophysics* **40**(3), 473–89.

VOZOFF, K. (1971) The effect of overburden on vertical component anomalies in AFMAG and VLF exploration: a computer model study. *Geophysics* **36**(1), 53–7.

VOZOFF, K. (1980) Electromagnetic methods in applied geophysics. *Geophys. Surv.* **4**, 9–29.

WAIT, J. R. (1951) A conducting sphere in a time varying magnetic field. *Geophysics* **16**(5), 666–72.

WAIT, J. R. (1952) The cylindrical ore body in the presence of a cable carrying an oscillating current. *Geophysics* **17**(2), 378–86.

WAIT, J. R. (1953) Induction by a horizontal magnetic dipole over a conducting homogeneous earth. *Trans. Am. Geophys. Union* **34**, 185–9.

WAIT, J. R. (1955) Mutual electromagnetic coupling of loops over a homogeneous ground. *Geophysics* **20**(3), 630–7.

WAIT, J. R. (1958) Induction by an oscillating magnetic dipole over a two-layer ground. *Appl. Sci. Res. Sec. B* **7**, 73–80.

WARD, S. H. (1967) The electromagnetic method. In *Mining Geophysics*, Vol. 2, Society of Exploration Geophysicists, Tulsa, pp. 224–372.

WARD, S. H. (1971) Discussion on 'Evaluation of the measurement of induced electrical polarization with an inductive system'. *Geophysics* **36**(2), 427–9.

WARD, S. H. (1972) Mining geophysics: new techniques and concepts. *Am. Mining Congr. J.* **58**, 58–68.

WARD, S. H. (1979) Ground electromagnetic methods and base metals. In *Geophysics and Geochemistry in the Search for Metallic Ores*, Geol. Surv. Can., Econ. Geol. Rep. 31, pp. 45–62.

WARD, S. H., RYU, J., GLENN, W. E., HOHMANN, G. W., DEY, A. and SMITH, B. D. (1974a) Electromagnetic methods in conductive terrains. *Geoexploration* **12**, 121–83.

WARD, S. H., PRIDMORE, D. F., RIJO, L. and GLENN, W. E. (1974b) Multispectral electromagnetic exploration for sulfides. *Geophysics* **39**(5), 666–82.

WARD, S. H., SMITH, B. D., GLENN, W. E., RIJO, L. and INMAN JR, J. R. (1976) Statistical evaluation of electrical sounding methods. Part II: Applied electromagnetic depth sounding. *Geophysics* **41**(6A), 1222–35.

WARD, S. H., CAMPBELL, R., CORBETT, J. D., HOHMANN, G. W., MOSS, C. K. and WRIGHT, P. M. (1977) Geophysics applied to detection and delineation of non-energy non-renewable resources. Report on grant AER76–80802, National Science Foundation; Dept of Geology and Geophysics, University of Utah.

WHITELEY, R. J. (ed.) (1981) *Geophysical Case Study of the Woodlawn Orebody, New South Wales, Australia*, Pergamon Press, New York, 588 pp.

WILLIAMS, D. A., STRANGL, R. O., SCOTT, W. J. and DYCK, A. V. (1973) Cavendish test range drilling program. Open file report of the Geological Survey of Canada.

WOLF, A. (1946) Electric field of an oscillating dipole over the surface of a two-layer earth. *Geophysics* 2(4), 518–34.

WON, J. (1980) A wide-band electromagnetic exploration method: some theoretical and experimental results. *Geophysics* 45(5), 928–40.

ZONGE, K. L. (1973) Minicomputer used in mineral exploration, or back-packing a box full of bits into the bush. Presented at the 11th Symposium on Computer Applications in the Mineral Industry, University of Arizona, Tucson.

FURTHER READING

AL'PIN, L. M., BERDICHEVSKII, M. N., VEDRINTSEV, G. A. and ZAGARMISTR, A. M. (1966) *Dipole Methods for Measuring Earth Conductivity*, selected and translated by G. V. Keller, Consultants Bureau, New York, 302 pp.

ANDERSON, W. L. (1977) Marquardt inversion of vertical magnetic field measurements from a grounded wire source. US Geol. Surv. Rep. GD–77–003, NTIS, Springfield, Va.

BACKUS, G. and GILBERT, F. (1970)Uniqueness in the inversion of inaccurate gross earth data. *Phil. Trans. R. Soc. London Ser A* 266, 123–92.

BANOS, A. (1966) *Dipole Radiation in the Presence of a Conducting Half-Space*, Pergamon Press, New York.

BHATTACHARYYA, B. K. (1964) Electromagnetic fields of a small loop antenna on the surface of a polarizable medium. *Geophysics* 29(5), 814–31.

BHATTACHARYYA, B. K. and PATRA, H. P. (1968) *Direct Current Geoelectric Sounding*, Elsevier, New York, 135 pp.

DEY, A. and MORRISON, H. F. (1973) Electromagnetic coupling in frequency- and time-domain induced polarization surveys over a multilayered earth. *Geophysics* 38(2), 380–405.

DIAS, C. A. (1968) A non-grounded method for measuring induced electrical polarization and conductivity. Ph.D. thesis, University of California, Berkeley.

DIETER, K., PATERSON, N. R. and GRANT, F. S. (1969) IP and resistivity type curves for three-dimensional bodies. *Geophysics* 34, 615–32.

FOUNTAIN, D. K. (1972) Geophysical case history of disseminated sulfide deposits in British Columbia. *Geophysics* 31(1), 142–59.

GEOEX PTY LTD (1977) Apparent resistivity time section, Willyama Complex, South Australia. Advertising brochure, Case-Study Series, No. 3.

GLENN, W. E., RYU, J., WARD, S.·H., PEEPLES, W. J. and PHILLIPS, R. J. (1973) Inversion of vertical magnetic dipole data over a layered structure. *Geophysics* 38(6), 1109–29.

HARRINGTON, R. F. (1968) *Field Computation by Moment Methods*, Macmillan, New York, 229 pp.

HOHMANN, G. W. (1973) Electromagnetic coupling between grounded wires at the surface of a two-layer earth. *Geophysics* 38(5), 854–63.

HOHMANN, G. W., KINTZINGER, P. R., VAN VOORHIS, G. D. and WARD, S. H. (1970) Evaluation of the measurement of induced polarization with an inductive system. *Geophysics* 35(5), 901–15.

JACKSON, D. D. (1972) Interpretation of inaccurate, insufficient and inconsistent data. *Geophys. J. R. Astron. Soc.* 28, 97–110.

KAUFFMAN, A. (1978a) Frequency and transient responses of electromagnetic field created by currents in confined conductors. *Geophysics* 43, 1002–10.

KAUFFMAN, A. (1978b) Resolving capabilities of the inductive methods of electroprospecting. *Geophysics* 43, 1392–8.

KELLER, G. V. and FRISCHKNECHT, F. C. (1966) *Electrical Methods in Geophysical Prospecting*, Pergamon Press, New York 517 pp.

KLEIN, J. D. and SHUEY, R. T. (1978) Nonlinear impedance of mineral-electrolyte interfaces, Parts I and II. *Geophysics* 43(6), 1222–49.

KUNETZ, G. (1966) *Principles of Direct Current Resistivity Prospecting*, Geoexploration Monograph Series 1, No. 2, Gebrüder Borntraeger, Berlin.

MCCRACKEN, K. G. and BUSELLI, G. (1978) Australian exploration geophysics: current performance and future prospects. Presented at 2nd Circum-Pacific Energy and Minerals Resources Conference, Honolulu.

MCNEILL, J. D. (1980a) Electrical conductivity of soils and rocks. Technical Note TN-5, Geonics, Ltd.

MCNEILL, J. D. (1980b) Electromagnetic terrain conductivity measurement at low induction numbers. Technical Note TN-6, Geonics, Ltd.

MADDEN, T. R. (1971) The resolving power of geoelectric measurements for delineating resistive zones within the crust. In *The Structure and Physical Properties of the Earth's Crust*, Ed. J. G. Heacock, American Geophysical Union, pp. 95–105.

MEYER, W. H. (1977) Computer modeling of electromagnetic prospecting methods. Ph.D. dissertation, University of California, Berkeley.

MILLETT JR, F. B. (1967). Electromagnetic coupling of colinear dipoles on a uniform half-space. In *Mining Geophysics*, Vol. 2, Society of Exploration Geophysicists, Tulsa, pp. 401–19.

MORRISON, H. F., DOLAN, W. and DEY, A. (1976) Earth conductivity determinations employing a single superconducting coil. *Geophysics* 41, 1184–1206.

MOTTER, J. W. (1980) Applications of formal search theory to exploration for non-fuel, non-renewable natural resources. Unpublished MBA thesis, University of Nevada, Reno.

NABIGHIAN, M. N. (1979) Quasi-static transient response of a conducting half-space: an approximate representation. *Geophysics* 44, 1700–5.

OLDENBURG, D. W. (1979) One-dimensional inversion of natural source magnetotelluric observations. *Geophysics* 44, 1218–44.

PARRY, J. R. (1969) Integral equation formulations of scattering from two-dimensional inhomogeneities in a conductive earth. Unpublished Ph.D. thesis, University of California, Berkeley.

PELTON, W. H. (1977) Interpretation of induced polarization and resistivity data. Unpublished Ph.D. thesis, University of Utah.

QUINCY, E. A., DAVENPORT, W. H. and LINDSEY, T. E. (1974) Preliminary field results on a new transient induction system employing pseudo-noise signals. *IEEE Trans. Geosci. Electr.* GE12, 123–33.

RIJO, J. (1977) Modelling of electric and electromagnetic data. Unpublished Ph.D. thesis, University of Utah.

ROY, A. and APPARAO, A. (1971) Depth of investigation in direct current methods. *Geophysics* **36**(5), 943–59.

SCOTT, W. J. and FRASER, D. C. (1973) Drilling of EM anomalies caused by overburden. *Can. Inst. Mining Metall. Bull.* **66**(735), 72–7.

SMITH, B. D. and WARD, S. H. (1974) A short note on the computation of polarization ellipse parameters. *Geophysics* **39**(6), 867–9.

SPIES, B. R. (1976) The derivation of absolute units in electromagnetic scale modelling. *Geophysics* **41**, 1042–7.

SPIES, B. R. (1979) Scale model studies of a transient electromagnetic prospecting system using an interactive mini-computer. *IEEE Trans. Geosci. Electr.* **GE–17**, 25–33.

SPIES, B. R. (1980) Interpretation and design of time domain electromagnetic surveys in areas of conductive overburden. *Bull. Aust. Soc. Explor. Geophys.* **10**(3), 203–5.

SUMNER, J. S. (1976) *Principles of Induced Polarization for Geophysical Exploration*, Elsevier, New York, 277 pp.

SUNDE, E. D. (1979) *Earth Conduction Effects in Transmission Systems*, Van Nostrand, New York, 373 pp.

TELFORD, W. M., GELDART, L. B., SHERIFF, R. E. and KEYS, D. A. (1976) *Applied Geophysics*, Cambridge University Press, Cambridge, 860 pp.

TRIPP, A. C., WARD, S. H., SILL, W. R., SWIFT JR, C. M. and PETRICK, W. R. (1978) Electromagnetic and Schlumberger resistivity in the Roosevelt Hot Springs KGRA. *Geophysics* **43**(7), 1450–69.

VAN NOSTRAND, R. G. and COOK, K. L. (1966) Interpretation of resistivity data. U.S. Geol. Surv., Prof. Paper 499, 310 pp.

WARD, S. H. (1971) Foreword and Introduction. In Special issue on electromagnetic scattering. *Geophysics* **36**(1), 1–8.

WARD, S. H., PARRY, W. T., NASH, W. P., SILL, W. R., COOK, K. L., SMITH, R. B., CHAPMAN, D. S., BROWN, F. H., WHELAN, J. A. and BOWMAN, J. R. (1978) A summary of the geology, geochemistry, and geophysics of the Roosevelt Hot Springs thermal area, Utah. *Geophysics* **43**, 1515–42.

WEIDELT, P. (1975) Electromagnetic induction in three-dimensional structures. *J. Geophys.* **41**, 85–109.

WIGGINS, R. A. (1972) The generalized inverse problem. *Rev. of Geophys. Space Phys.* **10**(1), 251–86.

ZONGE, K. L. and WYNN, J. C. (1975) EM coupling, its intrinsic value, its removal, and the cultural coupling problem. *Geophysics* **40**(5), 831–50.

Chapter 5

RADON MAPPING IN THE SEARCH FOR URANIUM

W. M. TELFORD

McGill University, Montreal, Quebec, Canada

SUMMARY

Instruments sensitive to radon emanation have been employed in uranium exploration, generally as a secondary survey method combined with gamma-ray detection. The technique appears to have an advantage over gamma-ray detectors because the depth of penetration may be considerably larger. On the other hand, the complicated process of radon diffusion from the source to the surface makes the accurate location of the source difficult. The survey technique is considered with regard to instrument development, procedures for field surveys and laboratory modelling and interpretation of data—which has been mainly qualitative in the past. A computer-adapted finite difference solution of equations for the diffusion and convection of radon through overburden is developed for a number of two- and three-dimensional models of simple geometry, which permits semi-quantitative interpretation of field anomalies. The complex nature of emanation and diffusion is stressed to illustrate the interpretation problem. A number of case histories are described to provide examples of the technique.

1. INTRODUCTION

Uranium exploration is in a relatively youthful stage, since it was introduced seriously only after the Second World War. Techniques include geological mapping, geochemistry, geophysics, exploration drilling and well logging. Although indirect geophysical methods— magnetics, IP, EM, etc.—have been used to some extent, the direct detection of radioactivity has been the main application, using instruments sensitive mainly to gamma radiation, occasionally to alpha and beta. The development of these has proceeded somewhat sporadically because of irregular demand for uranium and consequent lag in design of field equipment. Dodd (1977) has listed 14 uranium exploration methods, originally identified by an IAEA panel in 1972, of which 8 are geophysical (5γ-ray, 2 radon, 1 non-radiometric).

Since most of the uranium and thorium decay series elements emit γ-rays, which have a range of a few hundred metres in air, it is attractive to carry out reconnaissance with airborne and surface γ-ray detectors. The range of these natural γ-rays in solid materials, however, is considerably less than 1 m. Thus the source must lie at the surface. At present the only direct geophysical method potentially capable of penetrating the soil and subsoil is the measurement of radon emanation. Helium detection is another possibility; it was probably the first nuclear technique, dating back to the 1920s, but it has been employed in uranium exploration only very recently.

Element 86, radon, a radioactive noble gas, occurs as a decay product in all three radioactive series of ^{238}U, ^{235}U and ^{232}Th. Its location in these series and its characteristics are shown in Table 1. The ability of radon to escape from its radium parent at the source and migrate any appreciable distance through porous rock or overburden will be

TABLE 1

RADON ISOTOPES OF THE URANIUM AND THORIUM SERIES

Product No.	Isotope	Half-life	Radiation	Energies and abundance
1	^{238}U	4.51×10^9 y	α, SF	
6	^{226}Ra	1622 y	α, γ	$\alpha = 4.8\,\mathrm{MeV}(95\%)$, 4.6 MeV(4%)
7	^{222}Rn	3.82 d	α, γ	$\alpha = 4.6(99+)$, etc.
1	^{235}U	7.13×10^8 y	α, SF	
7	^{233}Ra	11.7 d	α, γ	$\alpha = 5.7(50)$, 5.6(24), etc.
8	^{219}Rn	4 s	α, γ	$\alpha = 6.8(82)$, 6.5(13), 6.4(5)
1	^{232}Th	1.39×10^{10} y	α, SF	
5	^{224}Ra	3.64 d	α	$\alpha = 5.7(95)$, 5.4(5)
6	^{220}Rn	51.5 s	α, γ	$\alpha = 6.3(99+)$, etc.

considered in Section 3. However, the longer half-life clearly favours ^{222}Rn over isotopes 220 and 219.

Since radon decays by α-emission, the detector monitors α-activity after collecting samples of soil air or water. In some instruments the collecting chamber walls are coated with a material which scintillates on α-bombardment, the scintillations subsequently being converted to voltage pulses. More recent devices detect the alphas as tracks on a film in the chamber (see Section 5 for equipment details).

2. HISTORICAL BACKGROUND

The isotope ^{220}Rn was discovered by Rutherford in about 1898. He called it thoron because of its place in the thorium series. Similarly ^{219}Rn, found by Debierne about the same time, was called actinon and ^{222}Rn, credited to F. E. Dorn in 1900, was first known as radium emanation. It was only later that all three were recognised as isotopes of element 86. The name radon was assigned in 1923 (Wahl and Bonner, 1951; Partington, 1957). It is the heaviest known gaseous element, with the highest melting point ($-71°$C), more soluble in water than the other inert gases and more soluble in organic solvents than in water, although dissolved salts decrease the solubility in water.

A survey of the literature on radon which is at all related to geophysics indicates that considerable theoretical and experimental work has been done on the mechanisms of emission from the source and diffusion through various materials, both lunar and terrestrial. In addition to reports of uranium exploration by radon detection, the possibilities for (i) mapping shallow structures, such as faults, (ii) earthquake prediction, (iii) monitoring radiation levels for environmental hazards and (iv) petroleum exploration have also been studied. The radon emanation technique appears to have been used widely in Russia both for uranium and oil exploration. Comprehensive reviews of all these subjects may be found in Tanner (1964, 1978).

3. GENERATION AND MIGRATION OF RADON ISOTOPES

3.1. Production of Radon by Uranium
Considering the radioactive decay process and assuming that equilibrium exists in the ^{238}U series, we have

$$\lambda_U N_U = \ldots \qquad \lambda_{Ra} N_{Ra} = \qquad \lambda_{Rn} N_{Rn} = \ldots$$

including all the offspring products to ^{206}Pb, where λ and N are the decay constant and number of atoms, respectively, for each member. Adding Avogadro's hypothesis, we can easily show that the mass of ^{226}Ra produced by 1 g of ^{238}U is 3.7×10^{-7} g. Since 1 curie (Ci) of ^{222}Rn is in equilibrium with 1 g of ^{226}Ra, we find that the activity of the radon daughter per gram of uranium is 3.7×10^5 •picocuries (pCi). Thus, by measuring the radon intensity we may determine the amount of uranium in the source. Of course if the series is not in equilibrium—at least for the first seven terms—the estimate will be too small.

In a field measurement, however, the fraction of radon which reaches the detector is enormously reduced unless the source outcrops. It is necessary to consider the attenuation between source and detector, which takes place during the initial escape and subsequent migration, the first involving recoil of the radon isotope from its radium parent and diffusion through the mineral grain, while the second is a diffusion and transport through permeable rock and soil. It is convenient to define an 'emanation coefficient', E, which is the ratio of radon atoms escaping to the total number created in a sample during the same time. Other equivalent terms found in the literature include 'emanating power', 'escape ratio', 'escape-to-production ratio' and 'percent emanation'.

3.2. Escape from the Source

Radon atoms produced by the parent radium have an initial recoil energy of 100 keV, corresponding to approximate ranges of 3×10^{-6} cm, 10^{-5} cm and $6-9 \times 10^{-3}$ cm in rock, water and air, respectively. They may be trapped in the source mineral grain or escape into a pore, depending on the grain dimension in the recoil direction. Furthermore, they may pass through the pore into an adjacent solid (e.g. another mineral grain), depending on the pore width and content with respect to the remaining kinetic energy. Atoms which remain in the pore constitute the direct recoil fraction of the emanation.

An additional possibility of escape has been discussed by several authors, involving an atom which may diffuse back into the pore from the adjacent grain in which it has created a pocket by vaporisation. This is known as the indirect recoil fraction.

Radon formed within the mineral grain at depths greater than the recoil range may escape by diffusion. It is not clear whether the diffusion fraction contributes significantly to the escape process: since diffusion goes on at a slower rate than recoil, it should discriminate against the short-life isotopes 220 and 219. The possibility of increased diffusion in zones of radiation damage has been investigated. The results are somewhat conflicting, although the intrusion of water in the mineral zone fractures may enhance both indirect recoil and diffusion fractions, by limiting the recoil range and inhibiting adsorption on the solid surface.

3.3. Emanation Coefficients

From this brief summary we can express the emanation coefficient in the form

$$E = E_{rd} + E_{ri} + E_d$$

the three terms representing direct and indirect recoil and diffusion, respectively. The first and third may be derived mathematically (Flügge and Zimens, 1939; Wahl and Bonner, 1951). For a direct recoil range R(cm) and a large grain with homogeneous distribution of radium, it can be shown that the radon fraction escaping by recoil is given by

$$E_{rd} = RA/4V$$

where A is the grain surface area (cm^2), V the grain volume (cm^3) and σ the density (g/cm^3). Using the specific surface area S which is $A/\sigma V$, this becomes

$$E_{rd} = R\sigma S/4$$

For atoms created at distances greater than R within the grain, the diffusion fraction is

$$E_d = \sigma S \sqrt{(D/\lambda)}$$

where D is the diffusion coefficient in the medium (cm^2/s) and λ the radon decay constant. Thus the emanation coefficient may be written

$$E = S\sigma[R/4 + \sqrt{(D/\lambda)}] \tag{1}$$

This model is oversimplified in several ways. It does not include the indirect recoil term, nor does it allow for depletion of surface layers by recoil enhancing diffusion rate; also, it assumes that the adjacent solid material is infinitely thick. For an appreciable diffusion fraction—say for ^{222}Rn—the value of D must be of the order of 10^{-20} in rock.

Changes in physical and chemical parameters affect the emanating power, although the relationships seem complex. For example, increasing water content in mineral crystal pores appears to increase emanation (Müller, 1930), but Starik and Melikova (1932) state that varying humidity works both ways. Emanation decreases with increasing grain size, although the exact relationship is not clear. There is a sharp increase at a definite temperature, but this is well above atmospheric range. Secondary uranium minerals (e.g. carnotite, autunite) have larger coefficients than most pitchblende and uraninite, probably because of their less compact structure. In general, however, the emanation in most minerals, rocks and soils is considerably higher than would be caused by recoil and diffusion alone. Tables 2 and 3 list values for some minerals, rocks and soils. Data from the USSR are from Starik and Melikova (1932), the remainder from Gilletti and Kulp (1955).

TABLE 2
EMANATION COEFFICIENTS OF MINERALS

Minerals and locale	$E(\%)$
Pitchblende (Sask.)	0.23 ± 0.02
Pitchblende (Sask.)	1.90 ± 0.09
Pitchblende (Sask.)	5.2 ± 0.3
Pitchblende (Sask.)	0.064 ± 0.008
Pitchblende (Idaho)	3.1 ± 0.4
Pitchblende (NWT)	6.7 ± 0.1
Pitchblende (Katanga)	1.61 ± 0.05
Pitchblende (Joachimsthal)	1.3 ± 0.1
Pitchblende (Mich.)	16.6 ± 0.3
Uraninite (N.C.)	0.58 ± 0.09
Samarskite (N.C.)	0.026 ± 0.008
Autunite (Aust.)	5.9 ± 0.3
Carnotite (Colo.)	17.3 ± 1.0
Carnotite (Colo.)	27.1 ± 0.8
Zircon (Brazil)	6.2 ± 0.1
Brannerite (Aust.)	6.0 ± 0.3
Apatite (USSR)	0.8
Magnetite (USSR)	4.0
Coffinite (USSR)	0.8
Sphene (USSR)	31.3
Eudeiolite (USSR)	8.8
Torbernite (USSR)	12–16.6
Carnotite (USSR)	32

TABLE 3
EMANATION COEFFICIENTS OF
ROCKS AND SOILS

Rock type	$E(\%)$
Gneiss	14 ± 0.8
Gneiss	1 ± 0.26
Granodiorite	16.9 ± 1.9
Granodiorite	40 ± 3.1
Granite	6.8 ± 0.9
Granite	32.7 ± 1.3
Pegmatite	4.3 ± 0.15
Syenite	9.3 ± 0.54
Basalt	2.5 ± 0.23
Gabbro	3.6 ± 0.58
Tuff	1.7 ± 0.12
Quartzite	5.3 ± 0.49
Sandstone	5.2 ± 0.22
Limestone	1.6 ± 0.42
Shale	2.6 ± 0.16
Soil type:	
Calcareous	33
Granitic	46
Sandy	9
Clay	30
Volcanic	49

3.4. Laboratory Measurements of Emanation

Soonawala (1976) carried out laboratory measurements of emanation coefficients on a variety of uranium mineral samples. The sample was first crushed and sieved to uniform size and placed in a bottle with a perforated cap at the bottom of a diffusion column (see Fig. 1). The latter consisted of a plastic pipe 125 cm long, 7.62 cm in diameter, the side walls pierced with 9 mm diameter holes at 15 cm intervals. These holes were sealed with silicone and the pipe filled with local soil to 120 cm.

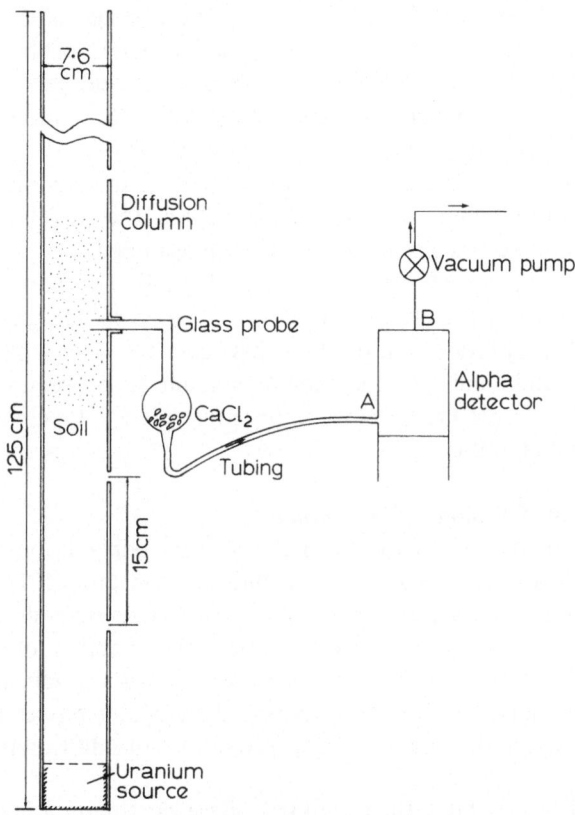

FIG. 1. Laboratory diffusion column experiment (schematic).

After standing for more than 14 days to bring the radon within 8% of equilibrium, the radon concentration was measured at various column heights by inserting a thin glass tube in the side hole. This tube was connected to the alpha detector, whose exit line contained a vacuum pump. As can be seen in Fig. 1, the input and output lines are equipped with valves A and B: with A closed and B open, the chamber is pumped to about 45 mm Hg pressure. With the valves reversed the detector is returned to atmospheric pressure, valve A is closed, and 5 successive 1-min counts recorded on the scaler attached to the alpha chamber. This routine permitted discrimination between ^{222}Rn and thoron. Two 1-min background counts were taken before each measurement sequence after flushing the system with air.

Subtracting average background from the average of 5 counts provided the radon concentration at the particular height of the column. On plotting corresponding values for different holes and extrapolating the logarithmic curve to zero distance, we obtain the source emanation concentration, while the slope of the curve determines the soil diffusion coefficient. Finally, after the uranium content of the sample was found, either by scintillometer or fluorimetric analysis, the correlation between emanation and source strength could be determined.

These measurements clearly confirmed that the radon emanation varies with the degree of consolidation of the rock. The output from a fine-grained graphitic argillite was particularly low, while a poorly compacted sandstone had the highest emanation. A preliminary laboratory study of this type, using representative rock types, is useful for interpreting field data.

3.5. Diffusion of Radon in Overburden

Subsequent migration of radon in the ground depends on several variables such as isotope decay rate, the diffusion constant of the pore-filling fluid, the fluid composition, phase and motion. For example water, while enhancing the emanation process, reduces the mobility of the thermal energy radon atoms because of its lower diffusion coefficient. On the other hand, migration may be increased if the water moves through the pores and carries the radon with it, providing an additional convection effect.

As would be expected, the radon gas phase existing in a liquid medium increases with temperature (e.g. for ^{222}Rn in water, 48% gas at 273 K, 84% at 313 K). At the same time the diffusion coefficient increases almost as the square of temperature, directly with pressure and more or less linearly with porosity. Diffusion is dominant in capillaries and small pores, while convection increases with the cross-section of the fluid medium. Since the relative contributions of the two mechanisms are rather difficult to determine, an 'effective' diffusion coefficient may be used (Culot et al., 1976).

Meteorological conditions affect radon migration, which generally increases with atmospheric temperature, inversely with pressure and rainfall, although mainly at shallow soil depths. Measurements on lunar samples returned with the Apollo missions indicated lower than terrestrial emanation due to fine-grained solids, low temperature and low pressure and moisture content.

3.6. Migration of Parent Isotopes

Tanner (1964) has noted that the mean migration distance for ^{222}Rn under optímum conditions is only 160 cm. Thus, the diffusion of radon alone cannot account for certain anomalies reported from field surveys where the source–detector separation is much larger. In addition to convection (Sections 3.5 and 4.2), it is necessary to consider the transport of certain of the radon parents such as the various radium isotopes, particularly ^{226}Ra, as well as ^{234}U and ^{238}U. For example, spring and surface waters, carrying finely dissolved radium and depositing it a considerable distance from the source, may produce a transplanted surface radon anomaly. The uranium isotopes may also contribute to the radon concentration in this fashion. Such migration would be characteristic of glacial terrain and regions where there is vertical movement of ground water.

This is a hydrogeochemical problem which complicates the interpretation of radon surveys, since the migration may take place laterally as well as vertically to produce misplaced anomalies. It has been well summarised by Smith et al. (1976).

4. THEORETICAL TREATMENT OF RADON DIFFUSION

Mathematical investigation of diffusion may be done analytically for simple one-dimensional structures and numerically for more complex geometry. In this section we develop analytical solutions for the infinite radon source covered with inactive overburden, the same situation allowing convection as well as diffusion and a layer of uniformly radioactive overburden. The treatment is the same as that of Alekseev et al. (1957).

4.1. Infinite Source

The geometry is shown in Fig. 2. Consider the elementary volume of area S and thickness dx at distance x above the plane source. For a radon concentration of N Ci/cm^3 in this space, Q and $Q+dQ$ Ci/s flux through

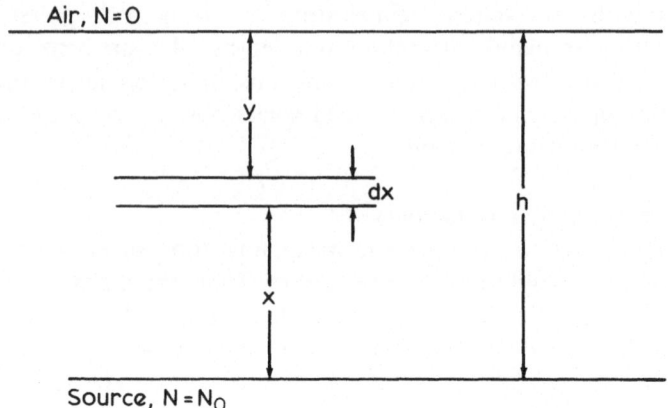

Fig. 2. Infinite source with barren overburden.

the lower and upper faces respectively, the rate of change of concentration within this volume is given by

$$\frac{\mathrm{d}}{\mathrm{d}t}(NS\,\mathrm{d}x)=Q-(Q+\mathrm{d}Q)-\lambda NS\,\mathrm{d}x=-(\mathrm{d}Q+\lambda NS\,\mathrm{d}x)$$

where λ is the radon decay constant. If D (cm^2/s) is the effective radon diffusion coefficient in the overburden,

$$Q=-D\frac{\mathrm{d}N}{\mathrm{d}x}S$$

Then for steady-state conditions $\mathrm{d}N/\mathrm{d}t=0$, and we have

$$\frac{\mathrm{d}^2N}{\mathrm{d}x^2}-\frac{\lambda N}{D}=0 \tag{2}$$

The general solution is $N=C_1\exp[-x\sqrt{(\lambda/D)}]+C_2\exp[x\sqrt{(\lambda/D)}]$ and for deep overburden we may say $N=N_0$ ($x=0$), $N=0$ ($x=\infty$), so that $N=N_0\exp[-x\sqrt{(\lambda/D)}]$.

On the other hand, if $N=0$ in the air at $x=h$, the solution is

$$N=\frac{N_0\sinh[(h-x)\sqrt{(\lambda/D)}]}{\sinh h\sqrt{(\lambda/D)}}=\frac{N_0\sinh y\sqrt{(\lambda/D)}}{\sinh h\sqrt{(\lambda/D)}} \tag{3}$$

This relation is plotted in Fig. 3 to show the attenuation of ^{222}Rn with depth of overburden for various values of diffusion coefficient.

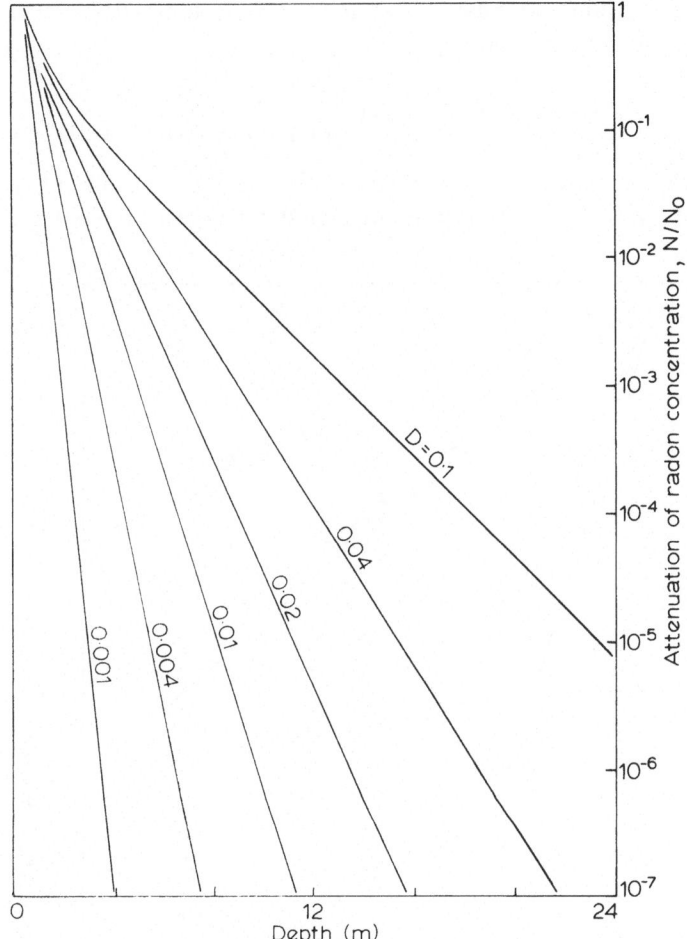

FIG. 3. Attenuation over an infinite source ($y = 50\,\text{cm}$).

4.2. Infinite Source with Convection

Equation (3) was derived on the assumption of a concentration gradient only; if the fluid medium (probably water) has a velocity v (cm/s), the original expression for Q is changed and

$$Q = -D\frac{dN\,S}{dx} + vNS$$

$$\frac{d^2N}{dx^2} - \frac{v\,dN}{D\,dx} - \frac{\lambda N}{D} = 0 \tag{4}$$

Using the same boundary conditions ($N = N_0$ at $x = 0$, $N = 0$ at $x = h$), we have for the solution

$$N = N_0 \exp[(h-y)v/2D] \frac{\sinh[y\sqrt{(v^2/4D^2 + \lambda/D)}]}{\sinh[h\sqrt{(v^2/4D^2 + \lambda/D)}]} \qquad (5)$$

Figure 4 shows the attenuation for various values of v when $D = 0.01\,\text{cm}^2/\text{s}$. The migration is considerably increased if v is appreciable.

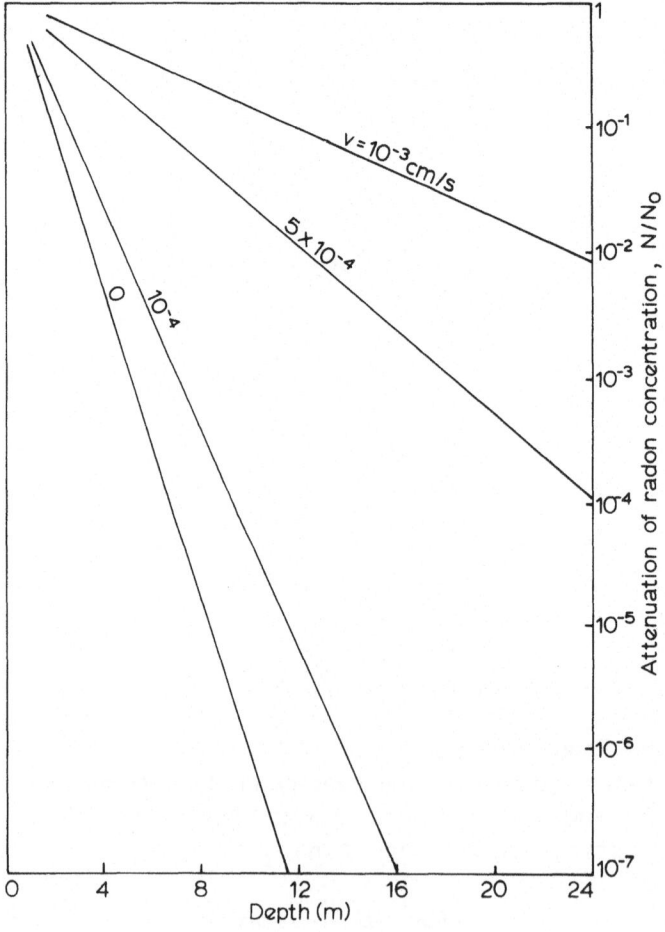

FIG. 4. Attenuation over an infinite source with fluid velocity in overburden ($y = 50$ cm, $D = 0.01$ cm^2/s).

For example, when $v = 10^{-3}$ cm/s $\simeq 1$ m/day, 10 cm/day and 1 cm/day, the factors are $\sim 10^5$, 45 and 1.6 respectively.

4.3. Homogeneous Radioactive Overburden

A soil or rock of density σ, porosity ϕ and radium concentration A' (g/g), whose emanation coefficient is a, generates $A'\sigma a$ Ci of radon in a volume ϕ of the pore space under equilibrium conditions. The emanation concentration in the pore space becomes

$$N = \frac{A'a\sigma}{\phi} \times 10^3 \text{ Ci/litre}$$

This condition prevails only where diffusion and convection are negligible. If the material contains 1 ppm uranium by weight, there will be 3.7×10^{-13} g of ^{226}Ra generated (Section 3.1) and we have

$$N = 370 \ a\sigma/\phi \text{ pCi/litre} \tag{6}$$

Taking extreme limits for a, σ and ϕ in rocks $(0.002 \leqslant a \leqslant 0.4$, $2.1 \leqslant \sigma \leqslant 2.8$, $0.004 \leqslant \phi \leqslant 0.34)$ and soils $(1.4 \leqslant \sigma \leqslant 1.93$, $0.39 \leqslant \phi \leqslant 0.9)$, we find that the overall range of N is about $1-10^4$. Correlation of radon concentration and uranium content over this range is illustrated in Fig. 5.

4.4. Numerical Techniques

Equation (2) is the so-called diffusion equation, an elliptical, partial differential form particularly amenable to solution by computer-controlled numerical techniques. For example, considered as a difference equation in two dimensions, the concentration at a point (x_0, y_0) in a homogeneous medium may be expressed in terms of the concentration values at four neighbouring points (McCracken and Dorn, 1964):

$$N(x_0, y_0) = (4 + \lambda k^2/D)^{-1} N(x_0 + k, y_0) + N(x_0 - k, y_0)$$
$$+ N(x_0, y_0 + k) + N(x_0, y_0 - k) \tag{7}$$

where k is very small but not zero. Similarly for three dimensions:

$$N(x_0, y_0, z_0) = [6 + \lambda k^2/D]^{-1} [N(x_0 + k, y_0, z_0) + N(x_0 - k, y_0, z_0)$$
$$+ N(x_0, y_0 + k, z_0) + N(x_0, y_0 - k, z_0)$$
$$+ N(x_0, y_0, z_0 + k) + N(x_0, y_0, z_0 - k)] \tag{8}$$

If either of these equation is satisfied for every point in the medium

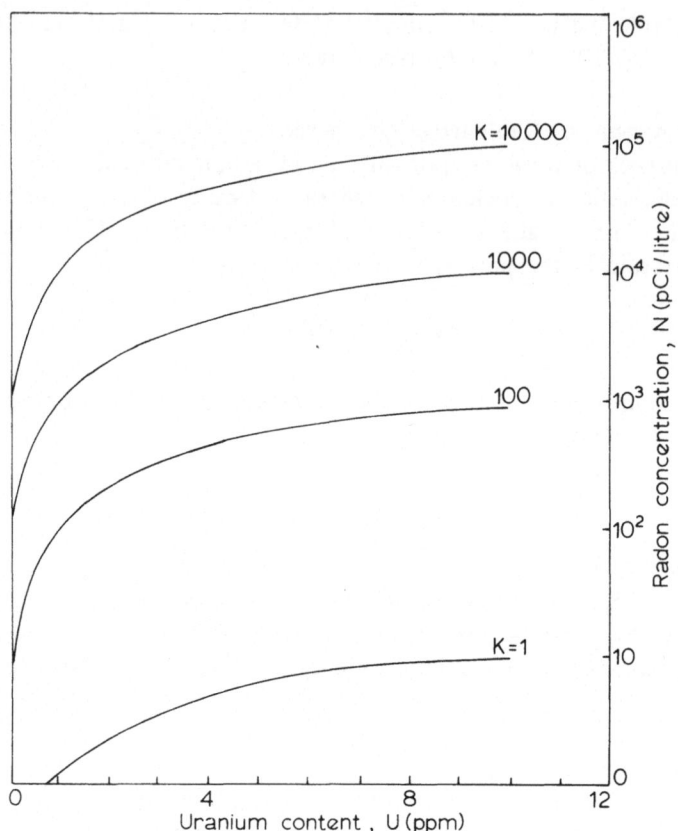

FIG. 5. Radon concentration as a function of uranium content. $N = KU$, $K = 370a\sigma/\phi$, where $a =$ emanation coefficient, $\sigma =$ density, $\phi =$ volume of pore space.

through which radon is diffusing and if all boundary conditions are satisfied, then the diffusion equation solution is obtained.

Several relaxation or iteration techniques are possible. In the Gauss–Seidel method a suitably spaced mesh is laid over the area (2-D) or volume (3-D) of interest and known values assigned to boundary mesh points. Initially N is assumed zero for all remaining points; with this complete value set, either eqn (7) or eqn (8) is solved in the computer, for a second iteration. Difference values between iterations are calculated and compared with a preselected maximum amount. Iteration is repeated until the solution has converged to a difference within the selected maximum.

Suitable mesh dimensions may be selected by trial and error, depending on the desired accuracy for a particular model and reasonable computer time for the analysis. For example, the infinite source solution obtained numerically was compared with the analytical results from eqn (3) for the following parameters: $y = 50$ and $100 \, \text{cm}$ (see Fig. 3), $h = 6 \, \text{m}$, $N_0 = 10^6 \, \text{pCi/litre}$, $D = 0.04 \, \text{cm}^2/\text{s}$. For $k = 100$, 50 and $25 \, \text{cm}$ the errors were about 6.6, 1.3 and 0.9%, respectively. Generally for dimensions appropriate to radon field work, a square mesh size of $50 \, \text{cm}$ was reasonably accurate without requiring excessive computer time; the mesh, however, need not be square.

4.5. Two-Dimensional Source

Figure 6 illustrates the geometry of a 2-D source of constant cross-section and strength, covered by barren overburden. The problem is to determine the concentration at depth y below the surface, equivalent to the detector position. Boundary conditions are $N = N_0$ at the source, $N = 0$ over the remaining source plane, over two vertical planes parallel to strike at a distance $5w$ either side of the source and in the air immediately above the surface. Profiles were computed (e.g. Fig 6 for $w = 200 \, \text{cm}$) over sources with $w = 30$, 200 and $1000 \, \text{cm}$ for various overburden depths and diffusion coefficients. Peak values over the source centres are shown in Figs 7, 8 and 9 for these conditions.

We may estimate a maximum depth attainable by this type of sampling. Assuming a 0.5% uranium concentration ($N_0 \cong 8 \times 10^4 \, \text{pCi/litre}$) and widths $w = 1 \, \text{m}$ and ∞, covered by overburden with $D = 0.04 \, \text{cm}^2/\text{s}$, the results are plotted in Fig. 10. This depth, of course, depends on the instrument sensitivity; if it were $1 \, \text{pCi/litre}$, the maximum is about 15 and $11 \, \text{m}$ of overburden for the infinite and $1 \, \text{m}$ source respectively. Recalling Section 4.2, this value may be greatly increased by convection in the overburden. The long stringer model is quite common in uranium exploration and the parameters used in the illustration are realistic.

4.6. Three-Dimensional Source

The geometry of a 3-D source is displayed in Fig. 11. Boundary conditions for the numerical solution are similar to those in Section 4.5. Figure 11 also shows the radon concentration over a square model with sides $1 \, \text{m}$, buried under $2 \, \text{m}$ of overburden with $D = 0.1 \, \text{cm}^2/\text{s}$. Minimum attenuation, directly over the centre for $y = 50 \, \text{cm}$, is 0.056. For the 2-D model with $w = 1 \, \text{m}$, the equivalent value was about 0.1. It may be seen

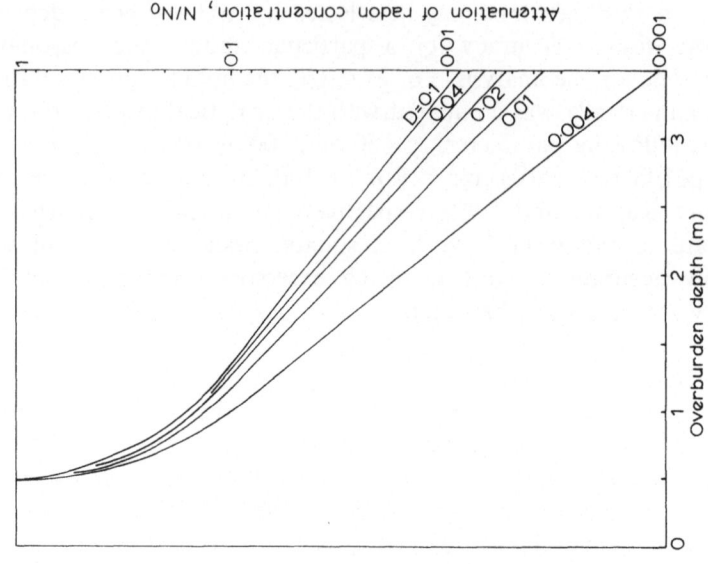

FIG. 7. Peak emanation over a thin two-dimensional source (width = 30 cm, y = 50 cm).

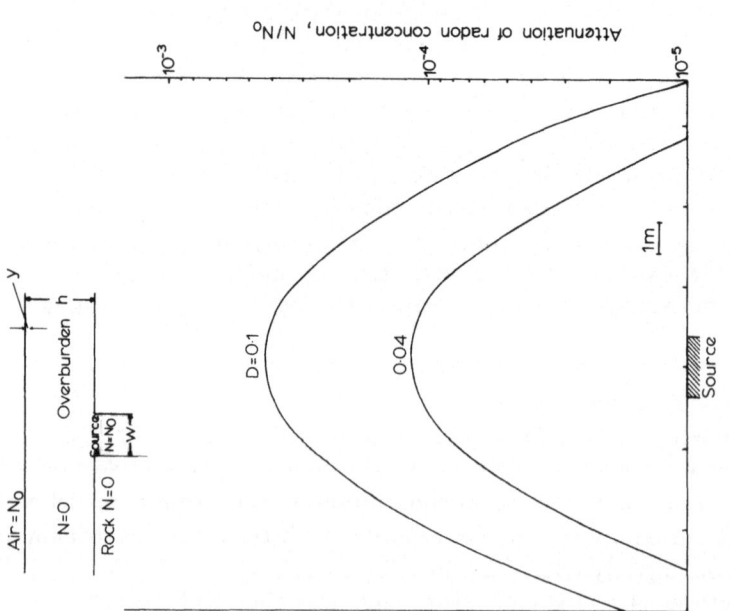

FIG. 6. Geometry and profiles for a two-dimensional source (width = 2 m, depth = 10 m, y = 50 cm).

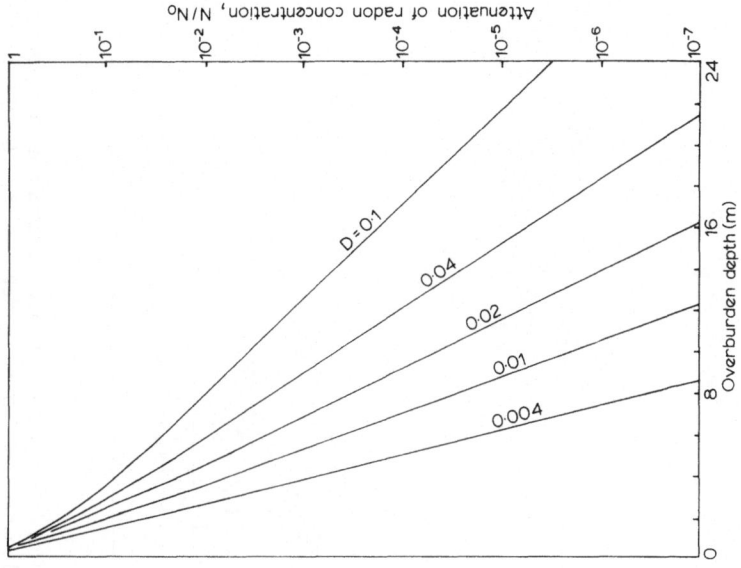

FIG. 9. Peak emanation over a thick two-dimensional source (width = 10 m, y = 50 cm).

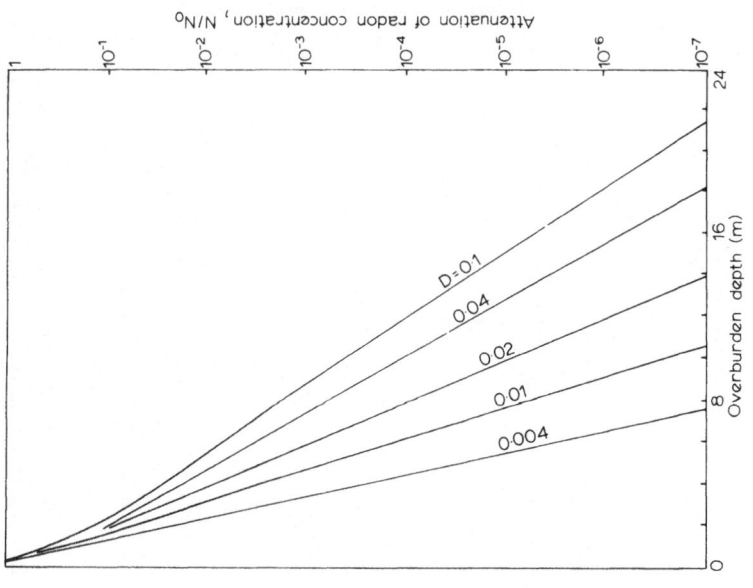

FIG. 8. Peak emanation over a two-dimensional source 2 m thick (y = 50 cm).

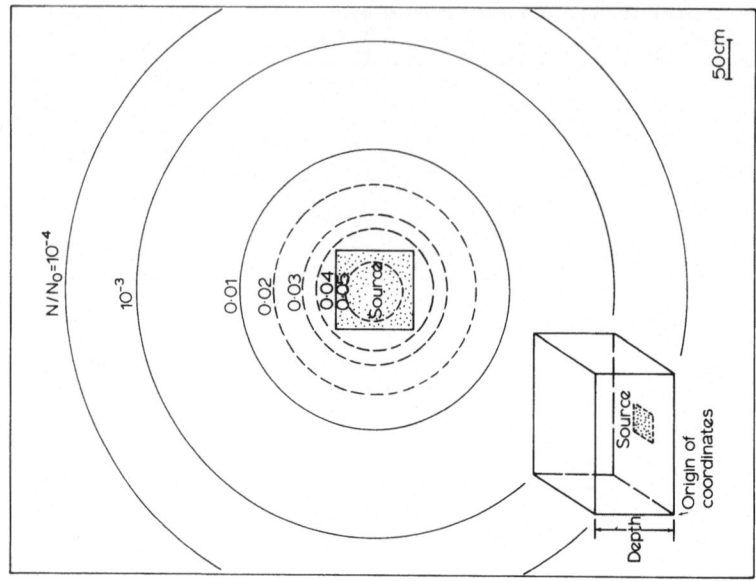

FIG. 11. Geometry and emanation for three-dimensional source (depth $= 2$ m, $D = 0.1$ cm^2/s, $y = 50$ cm).

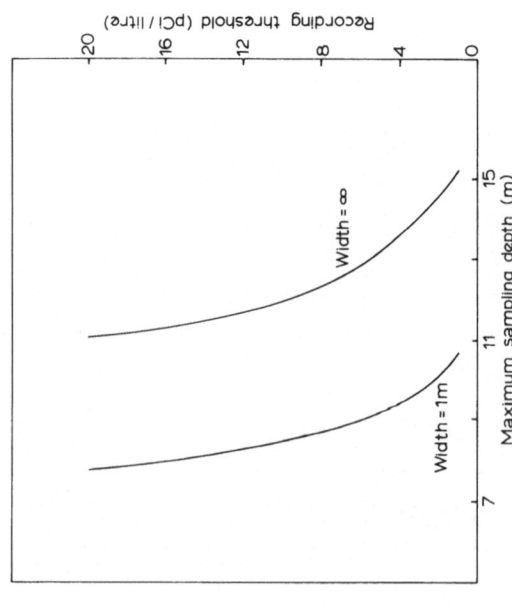

FIG. 10. Maximum depth penetration of emanometers. Radon concentration at source $N_0 = 8.25 \times 10^4$ pCi/litre; effective radon diffusion coefficient $D = 0.04$ cm^2/s.

from Fig. 11 that the off-centre response at a horizontal distance equivalent to the overburden depth has decreased by a factor of 5.6.

Another 3-D model, with dimensions $1 \times 5\,\text{m}$, produced almost the same results as the 2-D case with $w = 1\,\text{m}$, indicating that the larger dimension need be only 5 times the smaller to become a 2-D model.

4.7. Multi-Layer Model with 2-D Source

The geometry is shown in Fig. 12 for two layers of overburden with different diffusion rates, over a 2-D source. An additional boundary

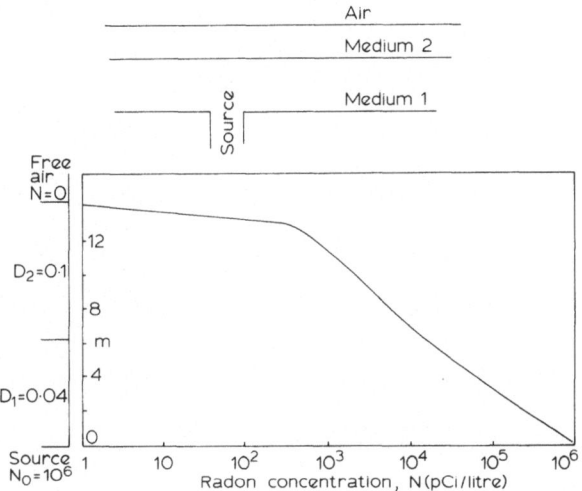

FIG. 12. Geometry and profile of vertical section through two-layered overburden.

condition is necessary at this interface, where the vertical radon flow (normal to the boundary) is continuous and we have

$$D_1(\mathrm{d}N/\mathrm{d}z)_1 = D_2(\mathrm{d}N/\mathrm{d}z)_2$$

that is, the product of diffusion coefficient and vertical emanation gradient in medium 1 must be equal to that in medium 2. A profile for a vertical section of this type is found in Fig. 12. The emanation decreases more rapidly in the lower medium because of its slower diffusion.

4.8. Two-Dimensional Fault

Figure 13 displays a 2-D fault with the fault plane, in which

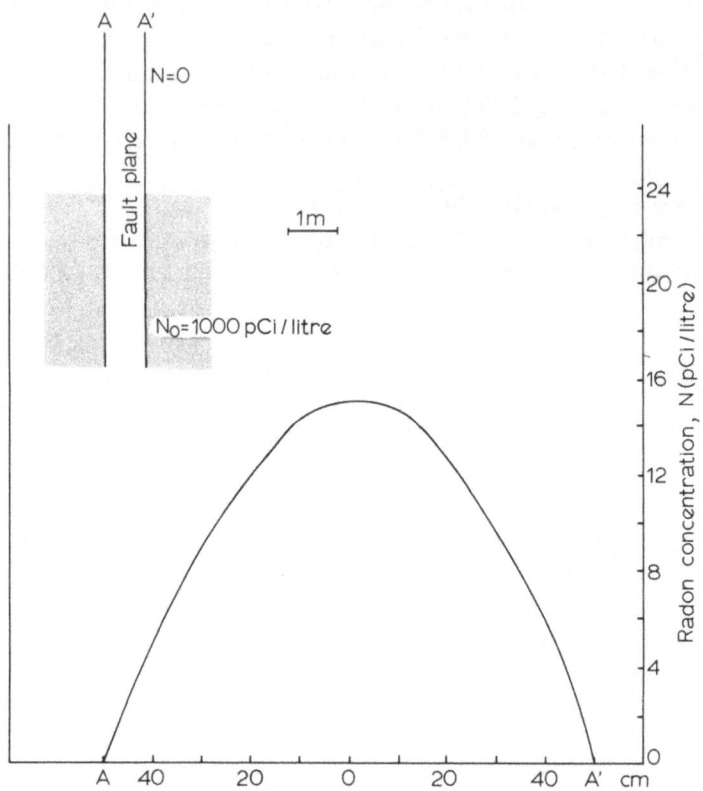

FIG. 13. Geometry and radon profile over a fault plane. Points A and A′ are fault plane boundaries.

$D = 0.1\,\text{cm}^2/\text{s}$, 1 m wide and 14 m deep. The lower 7 m section is radioactive with $N_0 = 1000\,\text{pCi/litre}$, the upper half barren. A horizontal profile at $y = 6\,\text{m}$ is shown.

4.9. Radon Diffusion Coefficient

The foregoing analysis emphasises the significance of the overburden diffusion coefficient for realistic interpretation of radon emanation data. There is a distinction between diffusion of radon in an isolated interstitial soil fluid (air, water, etc.) and diffusion for the entire material, the latter being a combination of solid- and fluid-filled pores, fractures and the like. Here we have used an effective diffusion coefficient (see Section 3.5), which is the value in the material of the interstitial spaces divided by the porosity of the whole medium.

Table 4 lists radon diffusion coefficients for several natural soils. These values are taken from Schroeder *et al.* (1965) and are in reasonable agreement with the data of Alekseev *et al.* (1957), who concluded that the range in soils of the USSR was between 0.04 and 0.004 cm^2/s. During the laboratory measurements described in Section 3.4 the local soil coefficient was found to be 0.03 cm^2/s.

TABLE 4

RADON DIFFUSION COEFFICIENTS OF NATURAL SOILS

Soil type	Soil condition	$D\,(cm^2/s)$
Unconsolidated glacial debris (Mass.)	Moist: matted grass, ground water	0.02
Consolidated sandstone (N. Mex.)	Moderately dry mine tunnel	0.03
Alluvium (Yucca Flat)	Dry, sandy	0.036
Alluvium (Yucca Flat)	Very dry, powdery, sparse cover	0.10

5. FIELD EQUIPMENT AND TECHNIQUES

As mentioned previously, most radon detectors are alpha monitors They may be classified as non-integrating or integrating types, depending on the time involved in detection and measurement. The short-period instrument is described first.

5.1. Radon Emanometer

Also known as 'sniffers', these devices collect diagnostic radiation for about 1 min. They have been described by Dyck (1968), Soonawala (1974) and Wollenberg (1977). The block diagram in Fig. 14 includes the following components: a tubular probe inserted in the soil, a pump to transfer soil air to the detector—which usually consists of a chamber whose inner walls are coated with silver-activated zinc sulphide—and a photomultiplier tube with count rate meter. (In one commercial model, however, the detector is a simple ionisation chamber.)

In performing field surveys the soil probe is inserted in a hole previously opened with a bar and hammer, generally about 50 cm deep. The solid spike bar is generally preferable to an auger, except for deep holes, since atmospheric air cannot enter the hole when the bar is in it, while the withdrawal of the bar sucks soil air into the hole. Soil air is

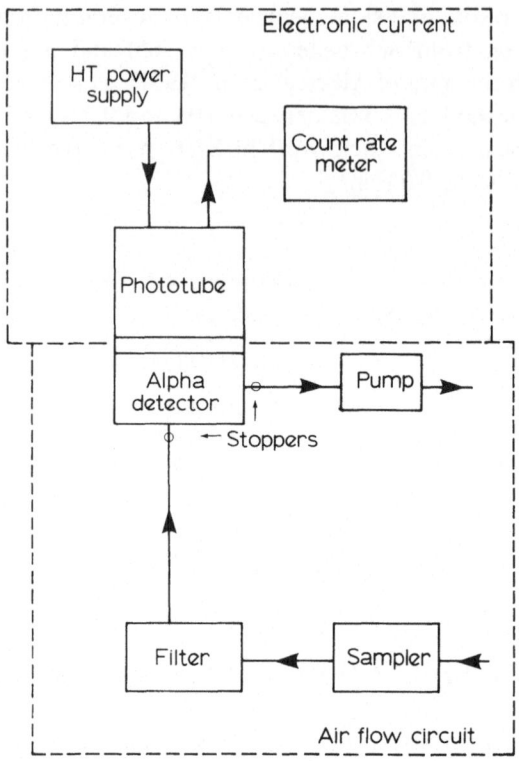

FIG. 14. Block diagram of pump emanometer.

transferred to the detector by the pump, where the radon-decay α-particles produce scintillations in the zinc sulphide. One wall of the chamber is transparent glass in contact with the photomultiplier tube, which converts scintillations to voltage pulses. These are fed to a digital scaler to produce counts for a preselected time.

Some filtering may be necessary in the pump line to remove dust and moisture. The detector chambers become contaminated from the build-up of repeated sampling, in spite of periodic flushing with atmospheric air. Consequently they must be changed frequently. In sampling soil water, air or inert gas is bubbled through the sample and collected in the detector chamber (Dyck, 1972).

It is important to distinguish between the short-lived thoron and ^{222}Rn. In this type of instrument such discrimination may be made qualitatively either by several successive 1-min counts or by retaining the air sample for several half-lives (~ 10 min). In the first measurement a sharp decrease will indicate the presence of thoron, the second reduces its relative response.

Another variation of the emanometer is the alpha probe, where the alpha detector is an integral part of the soil probe. Insertion in the soil circulates air past the sensor, which is a Perspex optical guide coated similarly to the previous model and connected to a phototube. This arrangement eliminates the pump, although it reduces the collection time.

5.2. Integrating Radon Monitor

This type of instrument collects and measures alpha activity at a station over a period of days or weeks. It is claimed that short-term variations in soil radon concentration are eliminated by the longer time of monitoring. Sections 3.4 and 3.5 made passing mention of the effects on emanating level due to atmospheric changes. Reports of diurnal and seasonal variations are conflicting, although emanation appears to decrease with (i) heavy rainfall, (ii) increasing barometric pressure, (iii) decreasing temperature and (iv) cold or frozen soil; the variations appear mainly small compared with mineralisation anomalies and the latter are generally repeatable over periods of months.

Three versions of the integrating instrument are described briefly. One was the Alpha Meter, which employed a probe similar to the emanometer, with a meter head. Total weight was about 1 kg. For radon surveys, 15 or 20 of these were planted at stations, picked up after reading the meter a couple of days later and moved to new stations.

The Track Etch employs a cellulose nitrate film taped on the inside bottom of a plastic cup, in which α-particles produce tracks which are later exposed by chemical etching. The inverted cup is placed in a soil hole for about 3 weeks, quite sufficient to average atmospheric effects while accumulating emanation. The cup walls are thick enough to exclude α-radiation from outside. In the first model there was no discrimination between thoron and radon emanation, but a later version has a membrane which delays passage of the soil air long enough to attenuate the 52 s thoron component while retaining most of the radon. After etching the amount of deposited radiation is determined from the track density, for example by viewing under a microscope.

The Alpha Cup uses a different type of detector, a silicon semiconductor which is fitted to the inside bottom of the container cup and attached by wire to an external electronic unit, the whole being buried in a soil hole. After about 3 days the electronic unit is retrieved and the α-

counts from the detector, now stored in its memory, are transferred to a reader. The detectors may be used indefinitely.

The Alpha Card is a slightly different version. A thin collecting membrane, mounted on a small card roughly the size of a 35 mm slide, has two silicon semiconductors attached either side of the card. These detect and store α-counts from particles passing through the membrane, providing improved counting geometry in the process. The card is hung vertically inside the inverted cup container and buried in the usual fashion, but only for about 12 h, after which it is removed and inserted in a portable card reader to measure the α-particle concentration. Correction for thoron may be obtained by a second sampling started a few hours later.

From these brief descriptions of radon detectors, it is evident that new instrument design has been slow. Scintillation detection is among the oldest of techniques used in nuclear physics, and there is nothing new in employing particle tracks. Radical changes in electronics, however, may be the main factor in future instrument development.

5.3. Field Operations

Surveys may be carried out by two men at a rate of about 70 stations/ day with the non-integrating emanometer. The usual station spacing may be 15 m on lines 50 m apart, decreasing to 7 m or less in anomalous areas. Following insertion of the probe into the soil and pumping to obtain a maximum reading, the chamber ports are closed and a count— generally for 1 min—initiated. If the reading is approximately background, which may normally be established from a few stations in a particular area, no further readings are required and the chamber is flushed for the next station. If the response is anomalous, 3 or 4 additional counts are taken, for the same period as the first, to achieve better counting statistics and to distinguish thoron from radon. Other data normally recorded should include hole depth, soil type and condition, and weather information such as wind, rainfall and temperature.

Using integrating instruments, the rate of planting detectors may be considerably faster than station occupancy with the emanometer, if a large supply is available. For example, it is stated that one man can lay out 30–60 Alpha Cards in an 8 h day. However, it is necessary to return later to obtain readings and retrieve the detectors for further measurements. Thus the survey is essentially slower than with the emanometer.

6. RADON SURVEYS AND CASE HISTORIES

In this final section we consider the results obtained in a variety of radon surveys carried out with different instruments over the past 12 years. Some of these were performed for test purposes, others for direct uranium exploration. Where possible, we include interpretation of field data and information gleaned from drilling, trenching, local geology, etc.

6.1. Water and Soil Gas Emanation Tests

Dyck (1969) reported measurements of radon concentrations in surface waters and soil gas at three sites: the Gatineau Hills, Quebec, and the Sudbury Basin and Elliot Lake areas, Ontario. These tests were carried out to assess the usefulness of the radon method in uranium prospecting. The sites correspond to a wide range of geological environment. For example, the main Sudbury mining area, rich in copper and nickel, has no known radioactive occurrences, although some have been found on the periphery. The Gatineau rocks contain uraninite, uranothorite and betafite in sub-economic quantities, while Elliot Lake has been a major producer of uranium, which occurs in quartz–pebble conglomerate.

In the Gatineau area several lakes were sampled along- and offshore—the latter at various depths—and at different seasons. The results indicated that radon in the waters was due to surface and underground drainage. In surface work, radioactive pegmatites under thin overburden were outlined more clearly by radon soil tests than by γ-ray detection; although radon levels varied by a factor of two during the field season, the anomalies were clearly reproduced.

Both radon and gamma response were negative across the Sudbury norite–granite contact associated with copper–nickel mineralisation and underground at Falconbridge. A reconnaissance survey in surface waters, however, showed variations in radon concentration in peripheral areas, and one lake anomaly north of Capreol appeared to have a similar origin to those in the Gatineau Hills.

Waters from lakes in the vicinity of Elliot Lake were sampled, two from the contaminated drainage channel south of the Quirke syncline, two in uncontaminated ground over the uranium-bearing Matinenda formation. The latter had radon levels an order of magnitude lower than the first two and lower than the lake waters in the Gatineau Hills. Thus, regional background levels are significant in resolving anomalies, as in other types of geophysical exploration. The ore zone at Rio Algom Quirke mine was more distinctly outlined by radon soil emanation than by γ-ray detection.

6.2. Weather Effects

Two sets of profiles over a narrow uraninite vein in contact-metamorphosed argillaceous rocks of southern Scotland, masked by glacial drift about 1 m thick, are shown in Fig. 15. These are from Bowie *et al.* (1971), who noted that there is no surface radiometric anomaly and, since uranium is not dispersed in the soil or subsoil in the vicinity of the vein, there would be no geochemical anomaly.

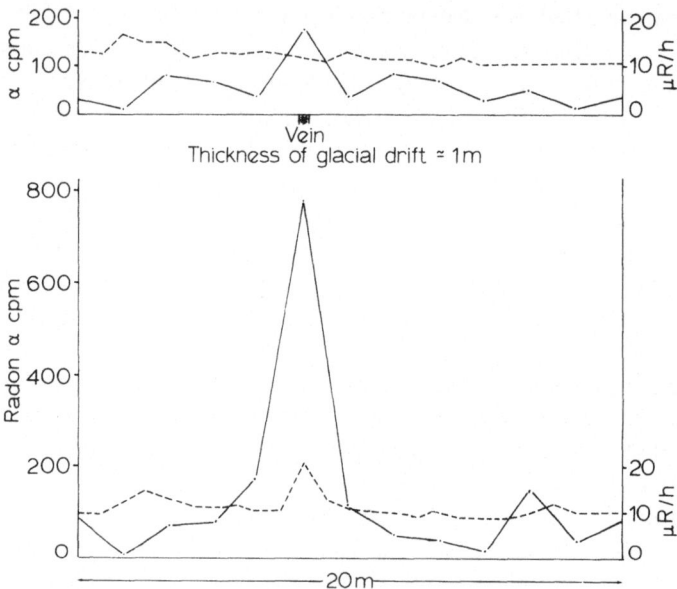

FIG. 15. Profiles over uranium-bearing vein in southern Scotland during dry weather (upper) and following heavy rain (lower). ----, surface radiometric; ———, soil radon. (After Bowie *et al.*, 1971.)

This example illustrates the atmospheric effect on emanometer response. The upper profile was obtained in dry weather, the lower following heavy rain. Although the dry weather anomaly is only about 20% of the other, the vein was clearly detectable in both cases.

6.3. Comparison of Radon and Gamma Response

Emanometer and scintillometer data from the Lisbon Valley area, Utah, taken from Stevens *et al.* (1971), are displayed in Fig. 16. Uranium deposits occur in the basal sandstone of the Triassic Chinle formation,

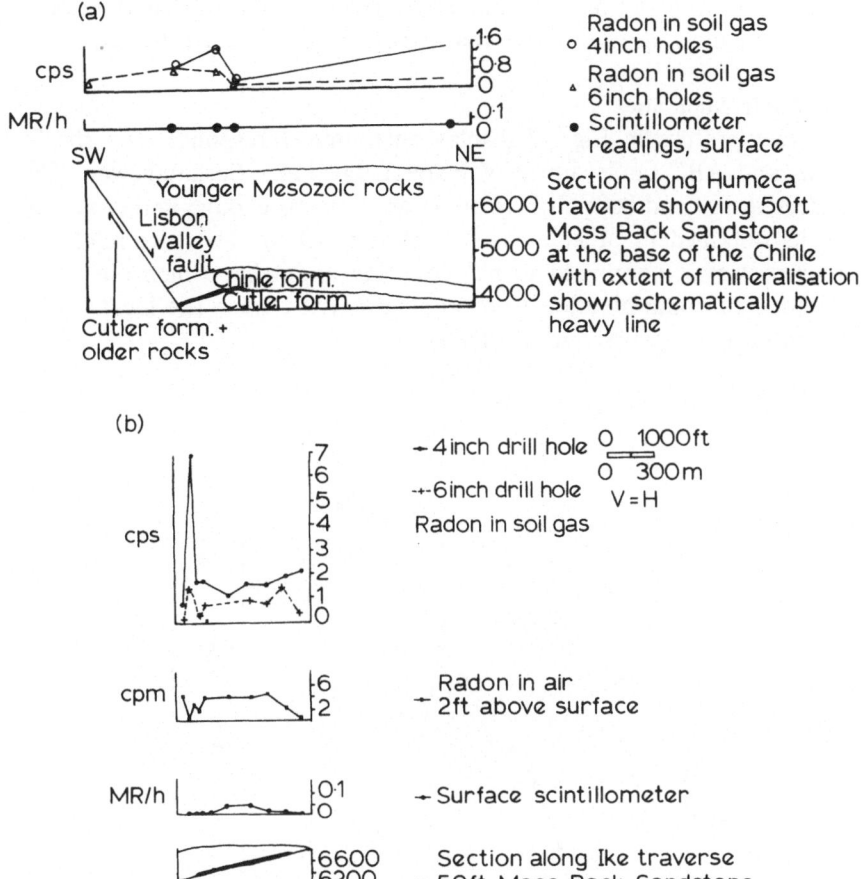

FIG. 16. Radon and scintillometer profiles, Lisbon Valley, Utah. (After Stevens
et al., 1971.)

resting unconformably on the underlying Permian Cutler formation,
located south-west of the north-west trending Lisbon Valley anticline,
which is broken by a normal fault.

Soil gas radon was collected at two depths, ~120 and 15 cm; the
former gave better response. Two sets of profiles with geological sections
are shown. Surface scintillometer readings were barren in the traverse of
Fig. 16(a). Although extremely small, the radon anomaly over the 2000 ft
(600 m) mineral section is most surprising. In Fig. 16(b) the sharp
response on the west end of the profile is thought to be contamination

from the nearby Ike mine dump. These profiles indicate that the radon measurement is more sensitive than γ-ray detection, but little else.

6.4. Fault Anomalies

The two profiles in Fig. 17 display emanometer response over the St Louis and ABC faults, north of Beaverlodge Lake, Saskatchewan. The respective 300 and 800 pCi/litre peaks are obvious enough and typical of results obtained in other areas (Soonawala, 1976). They are relatively small, however, compared to neighbouring anomalies which are probably due to contamination from mining activity, since the local U_3O_8 deposits are associated with the faults.

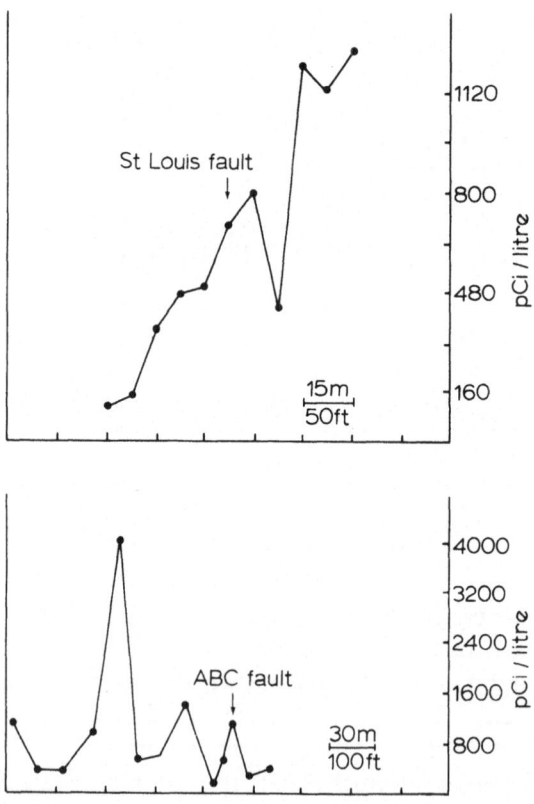

FIG. 17. Radon profile over two faults north of Beaverlodge Lake, Saskatchewan.

6.5. Cluff Lake Track Etch Survey

Beck and Gingrich (1976) describe a Track Etch orientation survey over the Cluff Lake 'N' deposit in northern Saskatchewan, which was done to assess the potential of this technique in glaciated terrain. The zone is in the form of several pitchblende-bearing lenses associated with the shallow dipping fracture sections in paragneiss and granites. Drilling outlined these zones at depths from 10 to 120 m. Some 160 Track Etch cups were buried in holes 70 cm deep on a grid which extended considerably beyond the known ore boundaries. The survey was repeated during winter with the cups at the bottom of the snow cover.

A contour map of the summer survey is shown in Fig. 18, together with an E–W geological section which crosses the zone but is not precisely tied to it. Track Etch readings more than 50 times background were obtained over the main orebody where the lens depths ranged from 10 to 90 m. The winter survey produced a similar contour pattern, although the absolute readings were lower; similar tests in areas of permafrost covering uranium mineralisation produced significant anomalies. It is also worth mentioning that the fracture zones contained between the faults F_1 and F_2 produce conductive anomalies with various electrical soundings, due to pyrite and graphite content.

6.6. Quantitative Interpretation of Field Data

Soonawala (1976) describes a variety of field surveys which have been interpreted by employing the models of Section 4. Diffusion column measurements of selected uranium-bearing samples from the survey areas provided a correlation between source strength and emanation.

6.6.1. Labrador

Two examples are considered. The uranium mineralisation occurs mainly as uraninite or pitchblendes disseminated through metasediments, in veinlets and thin seams along fractures, shears and faults and in pegmatites containing magnetite and sulphides.

The first radon profile is shown in Fig. 19(a). Here, the peak is about 8300 pCi/litre above background. A sample from a showing within 8 km of this survey produced 5.7×10^4 pCi/litre of ^{222}Rn per % U_3O_8, adjacent to the source, when measured in the laboratory. Actual mineralisation, exposed by stripping, was 1–1.5% U_3O_8 in a zone about 2 m wide at a depth of 2.1 m. From Fig. 8, showing peak emanation over a 2-D source 2 m wide, we find $N/N_0 \simeq 0.13$ for a depth of 2 m when $D = 0.04\,\mathrm{cm}^2/\mathrm{s}$ (although in the upper left area of the plot the value is not particularly

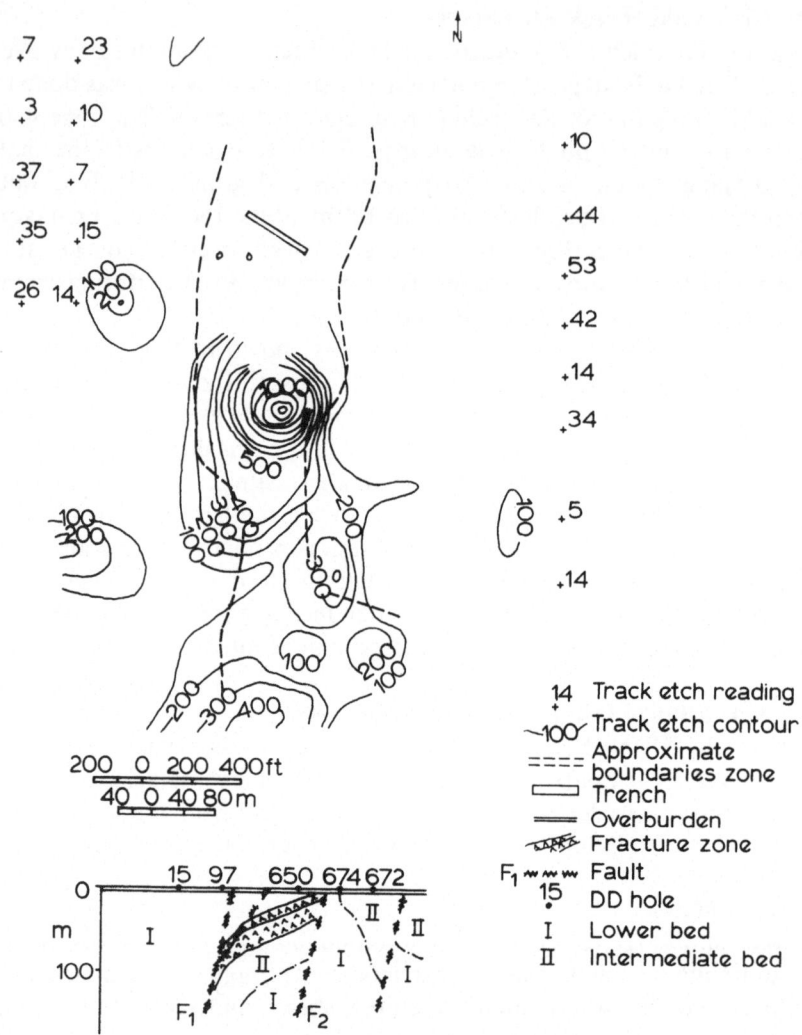

FIG. 18. Track Etch survey, Cluff Lake 'N' zone, and section. (After Beck and Gingrich, 1976.)

sensitive to variation in D). Thus, for 1.25% U_3O_8 the calculated emanation is $0.13 \times 7.1 \times 10^4 \simeq 9300\,\text{pCi/litre}$, which is within 10% of the measured value.

The second profile (Fig. 19(b)) has a peak value of ~ 2840 pCi/litre over background. Stripping revealed a zone of 0.05% U_3O_8, 3 m wide

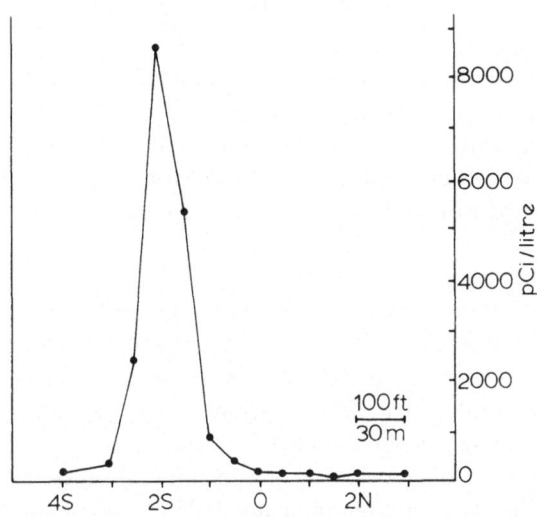

FIG. 19(a). Radon profile over Ribs Lake Showing, Labrador.

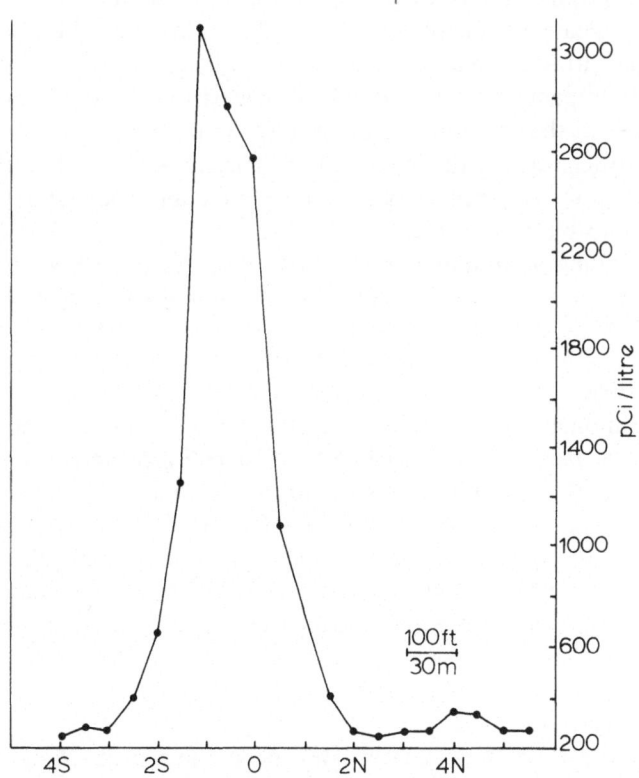

FIG. 19(b). Radon profile over Emben Prospect, Labrador.

under 1.2 m of overburden. From a 2-D plot of the width (not shown), $N/N_0 \simeq 0.41$ with $D = 0.04 \, \text{cm}^2/\text{s}$, therefore $N_0 = 2840/0.41 = 6875 \, \text{pCi}/$ litre. Compared with a laboratory sample collected some distance from the survey area, which assayed $1\% \, U_3O_8$ for $1.25 \times 10^5 \, \text{pCi/litre}$, i.e. 0.06% for 6875 pCi/litre, this agreement is also very good.

6.6.2. Saskatchewan

The area is Precambrian, with Athabaskan sediments and volcanics lying unconformably on Archean formations. There are numerous faults. Most of the ore-grade uranium is in veins, lenses and stringers of relatively high grade surrounded by dissemination halos, frequently in siliceous argillites; some mineralisation occurs in pegmatites.

A double peak profile over low-grade mineralisation in this area is illustrated in Fig. 20. Drill data indicated $0.05\% \, U_3O_8$ over 30 cm and an overburden 1–1.5 m thick. The larger peak at 13S is $N = 2680 \, \text{pCi/litre}$ above background. Laboratory measurements on samples from the area, however, indicate a value of $N_0 \simeq 3.8 \times 10^4 \, \text{pCi/litre per} \, \% \, U_3O_8$ and for 0.05% concentration this becomes 1900 pCi/litre, which is smaller than N. This discrepancy may be caused by at least two factors. Possibly the 0.05% grade at the drill intersection is lower than that immediately below the overburden and/or the emanation coefficient is larger than that used above, owing to the presence of secondary minerals. The first assumption is more plausible at this site.

In developing a computer model to fit the field profile, we must also account for the second peak at $13 + 30S$. Assuming a thin central mineral zone of 0.14%, 0.5 m wide, surrounded by a halo of 0.03%, all under 1 m overburden, a model profile may be fitted quite well to the field result, as is clear in Fig. 20.

The emanometer contours in Fig. 21 are from a survey over swampy ground. The peak on the baseline is 1000 pCi/litre above a very low background of 30 pCi/litre, suggesting a 3-D anomaly. The two percussion holes, however, were barren ($\leqslant 0.02\% \, U_3O_8$). This false anomaly may be due to contamination from an open pit operation about 100 m north, since the outer contours suggest a NE–SW strike. There may also be a minor depression at the baseline and an excess of carbonaceous material in the swamp.

6.6.3. Quebec

This example is from a survey over pegmatites near Mont Laurier, 250 km north of Montreal, where a 45° drill hole intersected 4.5 m of

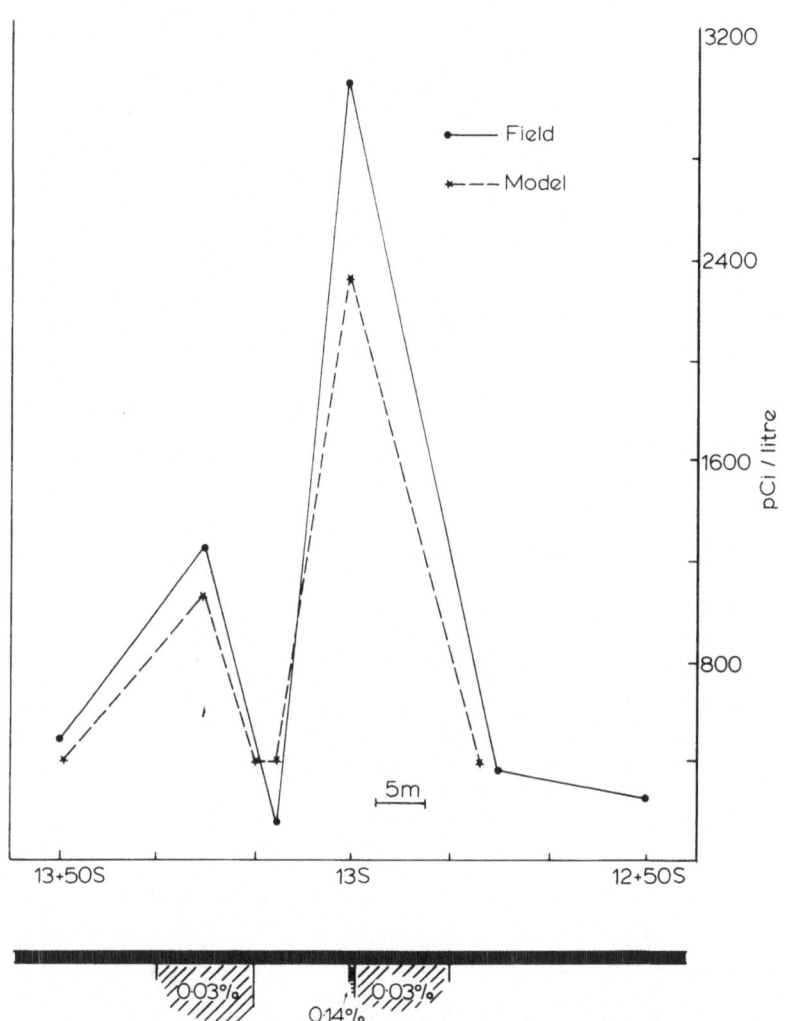

FIG. 20. Radon profile and computer model, Saskatchewan.

1.5% U_3O_8 approximately 21 m below the surface. The radon profile and drill location are shown in Fig. 22. Overburden depth was about 3 m. From Fig. 22 the peak response is 1.3×10^4 pCi/litre above background. Using curves of the type shown in Fig. 9 for $w = 5$ m and overburden 3 m, we obtain $N/N_0 = 0.08$. Thus $N_0 = 1.7 \times 10^5$ pCi/litre. Selecting a characteristic quartzo-feldspathic laboratory sample—in this case from Labrador—we find that 1% U_3O_8 corresponds to $N_0 = 1.2 \times 10^5$ pCi/litre.

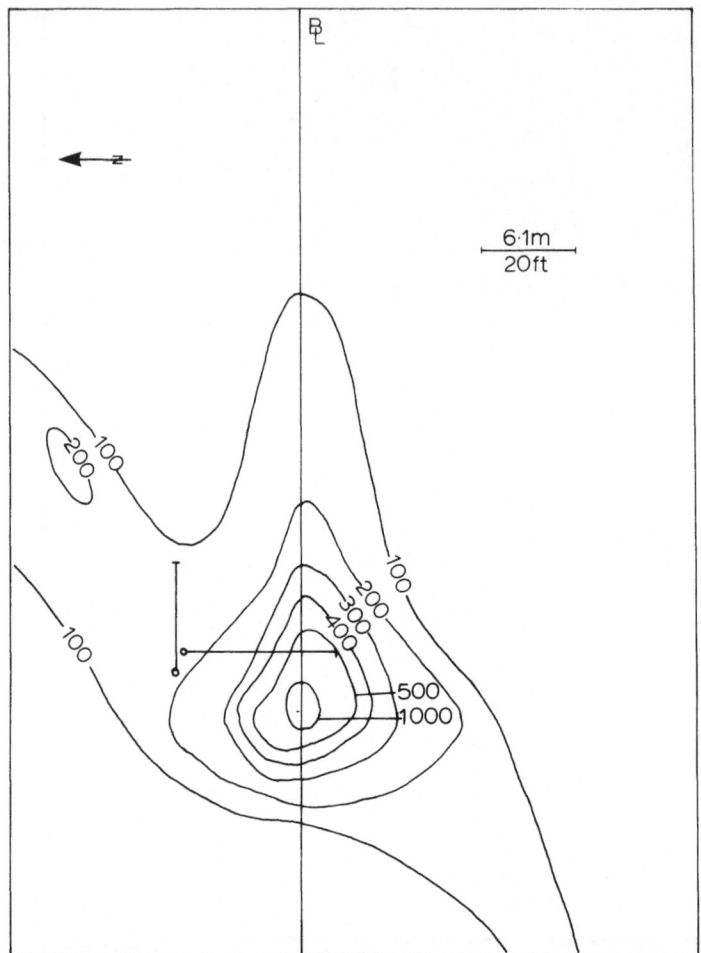

FIG. 21. Radon contours (pCi/litre) in swampy ground, Saskatchewan.

Hence 1.7×10^5 corresponds to $1.4\% \, U_3O_8$, in excellent agreement with drill data.

6.6.4. Deep Overburden
Caneer and Saum (1974) reported a case history where uranium ore overlain by about 115 m of overburden was detected by an emanometer survey. Their instrument readings are in counts per minute (cpm) and the sensitivity in terms of pCi/litre is not given. However, assuming that

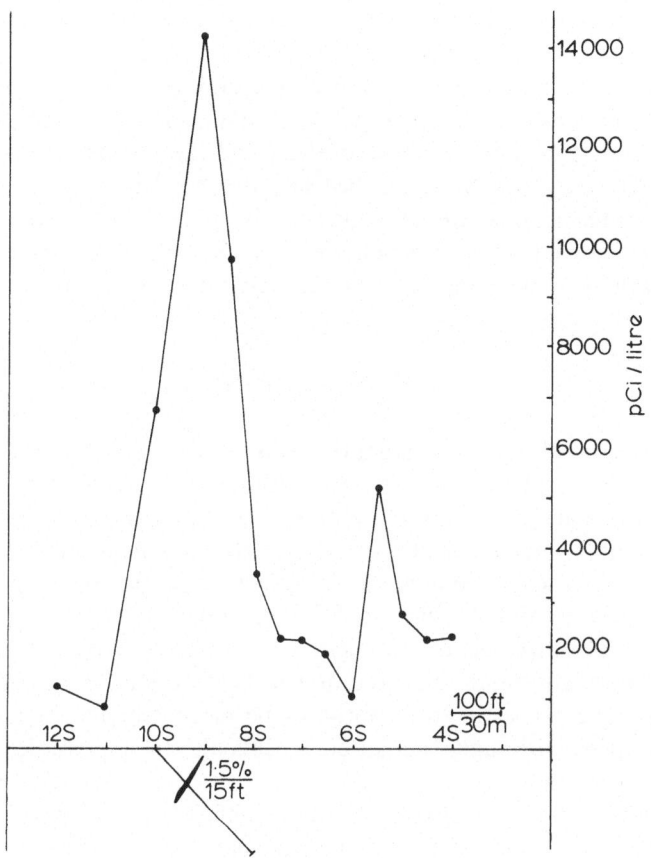

FIG. 22. Radon profile and drill section over uranium showing, Mont Laurier, Quebec.

1 cpm $\simeq 10$ pCi/litre, we can extract the following information from their report:

Background $\simeq 100$ pCi/litre; peak above background $\simeq 6000$ pCi/litre
Minimum depth of overburden $\simeq 107$ m; depth of detector $= 1$ m

Using eqn (3) for the infinite source, with $D = 0.1$ cm^2/s, we obtain $N/N_0 \cong 10^{-22}$ which, as one might expect, is highly unrealistic. From eqn (5), which includes overburden convection over the same source and assuming $v = 0.004$ cm/s $\simeq 3.5$ m/day, we find $N/N_0 \simeq 4 \times 10^{-3}$. Finally, assuming an effective emanation coefficient, typical of Wyoming sandstone, of 2×10^6 pCi/litre per % $U_3 O_8$, the grade is 0.73% $U_3 O_8$ for the source.

This is an extreme case, where convection is absolutely necessary unless the mineralisation has migrated through the overburden (Section 3.6). Note that in the first Saskatchewan and the Quebec examples, the model 2-D sources were at the bedrock–overburden interface, whereas the drill intersections are at considerably greater depth, that is, the source is assumed to have a definite depth extent. If this were not the case, it would be necessary to have convection in the rock section.

7. CONCLUSIONS

The detection of radon emanation is a useful, if somewhat limited, technique for uranium exploration. Although its maximum depth of penetration is generally considered to be no greater than 10–15 m, this is a considerable improvement over γ-ray detectors. Accurate location of sources may also be limited, as in geochemical prospecting, by the diffusion process and the high mobility of uranium itself.

This use of simple geometric models has been developed to the stage where semi-quantitative interpretation of field anomalies is possible. In this respect, depth estimates appear to be more reliable than grade of mineralisation, although the latter is difficult or impossible in most exploration methods.

It is clear that the effective emanation coefficient is a highly significant parameter affecting the radon technique, since it varies widely as a result of primary and secondary mineralisation, compaction, etc. For this reason a compilation of emanation coefficients of uranium-bearing rocks is very desirable, both for specific surveys, as was done for the examples in Section 6.6, and for general use.

The significance of convection in increasing the sensitivity of the radon method is also obvious. Soonawala (1976) has suggested pumping compressed air into drill holes or other entries to provide an artificial upward flow of interstitial fluid through surrounding rock and overburden.

ACKNOWLEDGEMENTS

The author is particularly indebted to Dr N. M. Soonawala, who provided much of the material for this work and reviewed the manuscript.

Thanks are also due to Dr Allan B. Tanner, US Geological Service, for material on radon migration and helpful review of a previous paper, and Dr Roger Lambert, Uranerz Exploration & Mining Ltd, for valuable discussion and comments.

REFERENCES

Because of the large number of references and to conserve some space, particularly pertaining to material in Sections 3 and 4, assignment of specific topics to their authors within the text has frequently been omitted. Instead these have been grouped under section headings.

Text References

ALEKSEEV, V. V., GRAMMAKOV, A. G., NIKANOV, A. I. and TAFEEV, G. P. (1957) Field emanation method. In *Radiometric Methods in the Prospecting and Exploration of Uranium Ores*, USAEC AEC-tr-3738 (Book 2), pp. 425–86.

BECK, H. L. and GINGRICH, J. E. (1976) Track Etch orientation survey in the Cluff Lake area, northern Saskatchewan. *Can. Inst. Mining Bull.* **69**(769), 104–9.

BOWIE, S. H. U., BALL, T. K. and OSTLE, D. (1971) Geochemical methods in the detection of hidden uranium deposits. In *Geochemical Exploration*, Proc. Int. Geochem. Explor. Soc., Spec. Vol. 11, Ed. R. W. Boyle, Canadian Institute of Mining and Metallurgy, Montreal, pp. 103–11.

CANEER, W. T. and SAUM, N. M. (1974) Radon emanometry in uranium exploration. *Mining Eng.* **26**(6), 28–9.

CULOT, M. V. J., SCHIAGER, K. J. and OLSEN, H. G. (1976) Prediction of increased gamma fields after application of a radon barrier on concrete surfaces. *Health Phys.* **30**(6), 471–8.

DODD, P. H. (1977) Uranium exploration technology. In *Geology, Mining and Extraction Processing of Uranium*, Ed. M. J. Jones, Institute of Mining and Metallurgy Symposium, London, p. 158.

DYCK, W. (1968) Radon-222 emanations from a uranium deposit. *Econ. Geol.* **63**(3), 288–9.

DYCK, W. (1969) Development of uranium exploration methods using radon. Geol. Surv. Can. Paper 69–46, Ottawa.

DYCK, W. (1972) Radon method of prospecting in Canada. In *Uranium Prospecting Handbook*, Ed. S. H. U. Bowie, M. Davis and D. Ostle, Institute of Mining and Metallurgy, London, pp. 212–41.

FLÜGGE, S. and ZIMENS, K. E. (1939) Determination of grain size and diffusion constant by the emanating power (theory of the emanation method). *Z. Phys. Chem. B* **42**, 179–220 (in German).

GILLETTI, E. J. and KULP, J. L. (1955) Radon leakage from radioactive materials. *Am. Mineral.* **40**(5–6), 481–96.

McCRACKEN, D. D. and DORN, W. S. (1964) *Numerical Methods and Fortran Programming*, Wiley, New York, pp. 365–92.

MÜLLER, H. Z. (1930) Die Emaniermethode als Mittel zur Untersuchung oberflächenarmer Salze. Z. Phys. Chem. A **149**, 257.

PARTINGTON, J. R. (1957) Discovery of radon. Nature (London) **179** (4566), 912.

SCHROEDER, G. L., KRANER, H. W. and EVANS, R. D. (1965) Diffusion of radon in several naturally occurring soil types. J. Geophys. Res. **70**(2), 471–4.

SMITH, A. Y., BARRETTO, P. M. C. and POURNIS, S. (1976) Radon methods in uranium exploration. In Exploration for Uranium Ore Deposits, IAEA Rep. STI/PUB/434, Vienna, pp. 185–211.

SOONAWALA, N. M. (1974) Data processing techniques for the radon method of uranium exploration. Can. Inst. Mining Bull. **67**(74), 110–16.

SOONAWALA, N. M. (1976) Diffusion of radon-222 in overburden and its application to uranium exploration. Unpublished Ph.D. thesis, McGill University, Montreal.

STARIK, I. E. and MELIKOVA, O. S. (1932) Emanating power of minerals. In USSR Proceedings of All-Union Conference on Radioactivity [in Russian], USAEC Rep. AEC-tr-4498, 1962, pp. 206–61 (trans.).

STEVENS, D. N., ROUSE, G. E. and DE VOTO, R. H. (1971) Radon-222 in soil gas: three uranium explorations case histories in the western US. In Geochemical Exploration, Proc. Int. Geochem. Explor. Soc., Spec. Vol. 11, Ed. R. W. Boyle, Canadian Institute of Mining and Metallurgy, Montreal, pp. 258–64.

TANNER, A. B. (1964) Radon migration in the ground: a review. In The Natural Radiation Environment: Houston Symposium Proceedings, Ed. J. A. S. Adams and W. M. Lowder, University of Chicago Press, pp. 161–190 (see also pp. 253–76).

TANNER, A. B. (1978) Radon migration in the ground: a supplementary review. USGS Open-file Rep. 78–1050, and in Proceedings of 3rd International Symposium on Natural Radiation Environment, Houston, Texas.

WAHL, A. C. and BONNER, N. A. (1951) Radioactivity Applied to Chemistry, Wiley, New York.

WOLLENBERG, H. A. (1977) Radiometric methods. Chapter 2 in Nuclear Methods in Mineral Exploration, Ed. J. Morse, Elsevier, Amsterdam, pp. 5–36.

Section 3

ANDREWS, J. N. and WOOD, D. F. (1972) Mechanism of radon release in rock matrices and entry into ground waters. Trans. Inst. Mining Metall. (London) Sec. B **81**, 198–209.

AUSTIN, S. R. (1975) A laboratory study of radon emanation from domestic uranium ores. In Radon in Uranium Mining, IAEA Rep. STI/PUB/391, Vienna, pp. 151–63.

BARRETTO, P. M. C. (1975) Radon-222 emanation characteristic of rocks and minerals. In Radon in Uranium Mining, IAEA Rep. STI/PUB/391, Vienna, pp. 129–50.

CLEMENTS, W. E. and WILKENING, M. H. (1974) Atmospheric pressure effects on ^{222}Rn transport across the earth–air interface. J. Geophys. Res. **79**, 5025–9.

JETER, E. W., MARTIN, J. D. and SCHUTZ, D. F. (1977) The migration of gaseous radionuclides through soil overlying a uranium deposit: a modeling study. US Dept Energy Rep. GJBX-67(77).

KOVACH, E. M. (1944) An experimental study on the radon content of soil gas. Am. Geophys. Union Trans. **25**, 563–71.

LAMBERT, G. (1977) Accumulation and circulation of gaseous radon between lunar fines. *Phil. Trans. R. Soc. London Ser. A* **285**, 331–6.

LAMBERT, G. and BRISTEAU, P. (1973) Migration of radon atoms implanted in crystals by recoil energy. *J. Phys. Colloque C5* **34**, 137–8 [in French]

LAMBERT, G., BRISTEAU, P. and POLIAN, G. (1972) Evidence of weakness of radon migration in interior of rock grains. *C. R. Acad. Sci. Paris Ser. D* **274**, 3333–6 [in French].

PEARSON, J. E. and JONES, G. E. (1965) Emanation of radon-222 from soils and its use as a tracer. *J. Geophys. Res.* **70**(20), 5279–90.

WILKENING, M. H. and HAND, J. E. (1960) Radon flux at the earth–air interface. *J. Geophys. Res.* **65**(10), 3367–70.

Section 4

BULASHEVICH, Y. P. and KHAYRITDINOV, R. K. (1959) On the theory of emanation diffusion in porous media. *Bull. Acad. Sci. USSR Geophys. Ser.* 1959, Pergamon, New York, pp. 1252–5.

GRAMMAKOV, A. G., KVASHNEVSHAYA, N. V., NOKONOV, A. I. and SOKOLOV, M. M. (1958) Some theoretical and methodological problems of radiometric prospecting and survey. In *Proceedings of 2nd UN Conference on Peaceful Uses of Atomic Energy*, Vol. 2, pp. 732–43.

QUET, C. (1975) Recoil emanating power and specific surface area of solids labelled by radium recoil atoms. 1. Theory for single solid particles. *Radiochem. Radioanal. Lett.* **23**(5–6), 359–68.

QUET, C., ROUSSEAU-VIOLET, J. and BUSSIÈRE, P. (1972) Recoil emanating power of isolated particles of finely divided solids. *Radiochem. Radioanal. Lett.* **9**(1), 9–18 [in French].

ROBERTSON, J. B. (1969) Diffusion from a gaseous source in a porous medium: a field and theoretical comparison. US Geol. Serv. Prof. Paper 650-D, p. D265.

Section 5

GAUCHER, E. (1976) Alpha meters: uranium prospecting by radon detection. *Can. Mining J.* **97**, 28–34.

MORSE, R. H. (1976) Radon counters in uranium exploration. In *Exploration for Uranium Ore Deposits*, IAEA Rep. STI/PUB/434, Vienna, pp. 229–39.

PEACOCK, J. D. and WILLIAMSON, R. (1961) Radon determination as a prospecting technique. *Trans. Inst. Mining Metall. (London)* **71**, 75–85; discussion pp. 271–7, 497–8.

PRADEL, J. (1956) Prospecting for uranium by radon. *Bull. Inf. Sci. Tech., Centre d'Etudes Nucléaires de Saclay* **9**(8), 1–8 [in French].

WARREN, R. K. (1977) Recent advances in uranium exploration with electronic alpha cups. *Geophysics* **42**, 982–9.

FURTHER READING

ARMSTRONG, F. E. (1972) Radiometrics proposed for exploration. Part 2. *Oil Gas J.* **70**(24), 152–61.

ARMSTRONG, F. E. and HEEMSTRA, R. J. (1972) Radiometrics proposed for exploration. Part 1. *Oil Gas J.* **70**(23), 88–91.

BECK, H. L. (1974) Gamma radiation from radon daughters in the atmosphere. *J. Geophys. Res.* **79**(15), 2215–21.

BELL, K. G., GOODMAN, C. and WHITFHEAD, W. L. (1940) Radioactivity of sedimentary rocks and associated petroleum. *Bull. Am. Assoc. Petrol. Geol.* **34**(9), 1529–47.

CROZIER, W. D. (1969) Direct measurement of radon-220 (thoron) exhalation from the ground. *J. Geophys. Res.* **74**(17), 4199–4205.

GABELMAN, J. (1977) *Migration of Uranium and Thorium*, Vol. 3: *Exploration Significance, Studies in Uranium Geology*, American Association of Petroleum Geologists, Tulsa, Okla.

KING, C. Y. (1978) Radon emanation on San Andreas fault. *Nature (London)* **271**(5645), 516–9.

MOGRO-CAMPERO, A. and FLEISCHER, R. L. (1977) Subterrestrial fluid convection: a hypothesis for long distance migration of radon within the earth. *Earth Planet. Sci. Lett.* **34**, 321–5.

SOONAWALA, N. M. and TELFORD, W. M. (1980) Movement of radon in overburden. *Geophysics* **45**, 1297–1315.

INDEX